T0178354

Springer Topics in Signal Processing

Volume 13

Series editors

Jacob Benesty, Montreal, Canada
Walter Kellermann, Erlangen, Germany

More information about this series at http://www.springer.com/series/8109

Orhan Gazi

Understanding Digital
Signal Processing

 Springer

Orhan Gazi
Electronics and Communication Engineering
  Department
Çankaya University
Etimesgut/Ankara
Turkey

ISSN 1866-2609   ISSN 1866-2617   (electronic)
Springer Topics in Signal Processing
ISBN 978-981-13-5277-5   ISBN 978-981-10-4962-0   (eBook)
DOI 10.1007/978-981-10-4962-0

Printed on acid-free paper

This Springer imprint is published by Springer Nature
The registered company is Springer Nature Singapore Pte Ltd.
The registered company address is: 152 Beach Road, #21-01/04 Gateway East, Singapore 189721, Singapore

Preface

In this book, we tried to explain digital signal processing topics in detail. We paid attention to the simplicity of the explanation language. And we provided examples with increasing difficulty. The reader of this book should have some background about signals. If it is possible, the reader should learn fundamental concepts on signals and systems since, in this book, more attention is paid on digital signal processing concepts rather than continuous time signal processing topics. Hence, we assume that the reader has fundamental knowledge about all types of signals and transforms.

All the topics in this book are presented in an orderly manner. We tried to simplify the language of this book as possible as we can. We also provided original examples explaining the aim of the subjects studied in this book. Numerical examples are provided for the comprehension of the subjects. Unnecessary abundance of mathematical details is omitted for the simplicity of the presentation language. In addition, to indicate both continuous and digital time frequencies, we preferred to use the same parameter. We thought that using two different parameters mixes the students' mind and it is not necessarily needed.

This book includes four different chapters. And in these chapters, sampling of continuous time signals, multirate signal processing, discrete Fourier transform, and filter design concepts are covered. In sampling of continuous time signals and multirate signal processing chapters, we provided some original practical techniques to draw the spectrum of aliased signals. In discrete time Fourier transform chapter, well-designed numerical examples are provided to illustrate the operation of the fast Fourier transform algorithm. In filter design chapter, both analog and digital filter design techniques are explained in detail. For the analog filters, we also provided analog filter circuit design methods for the designed analog filter transfer function.

Maltepe/Ankara, Turkey
November 2016

Orhan Gazi

Contents

Chapter 1
Sampling of Continuous Time Signals

Signal is a physical phenomenon that carries information. This physical phenomenon is described by mathematical functions, and usually the signal and its mathematical function are used for one another, i.e., synonymous. For instance, when we talk about a sinusoidal signal, we use the sinusoidal function, a mathematical function, to characterize the signal, and the name sinusoidal is used for the signal. Signals are usually depicted in graphs to observe their behavior and analyze them. Sinusoidal signals are the main signals and all the other signals can be considered as being made up of sinusoidal signals with different frequencies and amplitudes. That is to say, any continuous time signal can be written as sum of sinusoidal signals with different frequencies and amplitudes. Rectangular signal, square pulse signal, impulse train signal, triangle signal can be given as examples of continuous time signals.

Digital signals are obtained from continuous time signals via sampling operation. Digital signals are represented as mathematical sequences, and the elements of these sequences are nothing but the amplitude values taken from continuous time signals at every multiple of the sampling period. Since in the last several decades a huge improvement is achieved at the development of the digital devices, it has become almost a must especially for electrical engineers to have a good knowledge of digital signals. Digital signals are almost available in every part of our life. Computers, TVs, speakers, mobile phones, house equipment, and most of the other electronic devices process digital signals. In this chapter, we discuss the construction of digital signals via sampling operation, their spectral analyses, the case of aliasing, and reconstruction of a continuous time signal from its samples.

© Springer Nature Singapore Pte Ltd. 2018
O. Gazi, *Understanding Digital Signal Processing*, Springer Topics
in Signal Processing 13, DOI 10.1007/978-981-10-4962-0_1

1.1 Sampling Operation for Continuous Time Signals

Let $x_c(t)$ be a continuous time signal. We take samples from the amplitudes of this signal at every multiple of T_s which is called sampling period and form a mathematical sequence. The obtained mathematical sequence is called digital signal.

The sampling operation is described by the formula

$$x[n] = x_c(nT_s) \quad n \in Z, \; T_s \in R \tag{1.1}$$

where n is of integer type and T_s is the sampling period.

The block diagram of the sampling operation is depicted in Fig. 1.1.

Let's now try to explain the sampling operation on a sinusoidal signal. The graph of the sinusoidal signal with period T is given in Fig. 1.2.

Let's now take some samples from the sine signal in Fig. 1.2, and within this purpose, let's choose sampling period as $T_s = \frac{T}{6}$. Samples from signal amplitude are taken at every multiple of T_s, and this operation is illustrated in Fig. 1.3.

The sampled amplitude values are placed into an array and expressed as a mathematical sequence. The mathematical sequence obtained from the above sampling operation can be written as

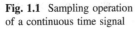

which is a digital signal obtained from a continuous time signal. The obtained mathematical sequence can also be displayed graphically as in Fig. 1.4.

Fig. 1.1 Sampling operation of a continuous time signal

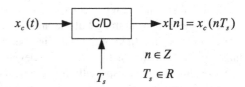

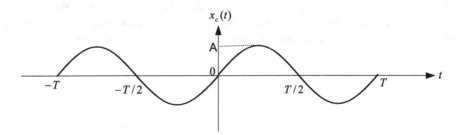

Fig. 1.2 Sine signal with period T

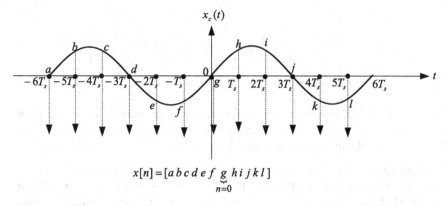

Fig. 1.3 Sampling of the sine signal

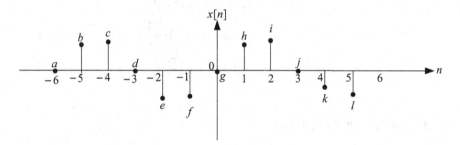

Fig. 1.4 Digital sine signal

If starting index value, i.e., $n = 0$, is not indicated in the mathematical sequence, the index of the first element is accepted as $n = 0$.

Graphical illustration is usually employed for easy understanding of the sampling operation and to interpret the meaning of the received signal. Let's consider the sampling of sine signal again and write a mathematical expression for the digital sine signal. The continuous time sinus signal with period T is written as

$$x_c(t) = \sin\left(\frac{2\pi}{T} t\right). \tag{1.2}$$

If the continuous time signal in (1.2) is sampled with sampling period $T_s = \frac{T}{6}$, we obtain the digital signal $x[n]$ whose mathematical expression can be calculated as

$$x[n] = x_c(t)|_{t=nT_s} \rightarrow x[n] = \sin\left(\frac{2\pi}{T} n \frac{T}{6}\right) \rightarrow x[n] = \sin\left(\frac{\pi}{3} n\right). \tag{1.3}$$

By giving negative and positive values to n we obtain the amplitude values of digital sine signal which can be shown as

$$x[n] = \left[\ldots \sin\left(-\frac{2\pi}{3}\right)\sin\left(-\frac{\pi}{3}\right)\underbrace{\sin\left(\frac{0\pi}{3}\right)}_{n=0}\sin\left(\frac{2\pi}{3}\right)\ldots\right]. \tag{1.4}$$

Example 1.1 Find the frequency and period of the continuous time signal $x_c(t) = \cos(2\pi t)$. Sample the given continuous time signal with sampling period $T_s = 1/8$ s and obtain the digital signal $x[n]$.

Solution 1.1 If $x_c(t) = \cos(2\pi t)$ is compared to the general form of cosine signal $\cos(2\pi f t)$, it is seen that the frequency of $x_c(t)$ is $f = 1$ Hz which can be used to find the period of the signal using $T = 1/f$ leading to $T = 1$ s. The sampling operation for $x_c(t) = \cos(2\pi t)$ with sampling period $T_s = 1/8$ s is done as

$$x[n] = x_c(t)|_{t=nT_s} \rightarrow x[n] = \cos(2\pi t)|_{t=nT_s}$$
$$\rightarrow x[n] = \cos\left(2\pi n\frac{1}{8}\right) \tag{1.5}$$
$$\rightarrow x[n] = \cos\left(\frac{\pi n}{4}\right).$$

1.1.1 Sampling Frequency

In communication theory; sampling frequency is one of the most important parameters. Sampling frequency is used more than sampling period. Sampling frequency shows the number of samples taken from a continuous time signal per-second. For this reason, it is an indicator of the quality of the continuous-to-digital converters. As sampling frequency increases more samples are taken per-second but this leads to an increase in transmission overhead.

As an example, if the sampling frequency is 1000 Hz i.e., 1 kHz, it means that every second, 1000 samples are taken from continuous time signal.

Verification
Let's now prove the above claim (the meaning of sampling frequency) for a continuous time periodic signal. Let $x_c(t)$ be a continuous time periodic signal, with period T and T_s be the sampling period. In this case, from one period of the signal a total of $\frac{T}{T_s}$ samples are collected. The continuous time period signal repeats itself $\frac{1}{T}$ times in 1 s. According to this information, in one second, the total number of samples taken from the signal equals to $\frac{T}{T_s} \times \frac{1}{T} \rightarrow \frac{1}{T_s}$ which is nothing but the sampling frequency.

Example 1.2 The continuous time signal $x_c(t) = \cos(2\pi ft)$ where $f = 1$ kHz is sampled with sampling frequency $f_s = 16$ kHz, and the digital signal $x[n] = x_c(nT_s)$ is obtained. According to the given information, find

(a) The number of samples taken from one period of the continuous time signal.
(b) The number of samples taken per-second from continuous time signal.

Solution 1.2 The number of samples taken per-second from continuous time signal equals the sampling frequency, i.e., $f_s = 16000$ samples are taken per-second. Since the period of the continuous time signal is $T = \frac{1}{1\ \text{kHz}} \rightarrow T = 1$ ms, the number of samples taken from one period of the signal is 16000×1 ms $\rightarrow 16$ samples.

1.1.2 Mathematical Characterization of the Sampling Operation

Impulse Train
Impulse train function is one of the most widely used mathematical expression appearing in sampling operation. For this reason, we will first inspect the impulse train function in details. The impulse train function is given as

$$s(t) = \sum_{n=-\infty}^{\infty} \delta(t - nT_s) \qquad (1.6)$$

where T_s is the sampling period. The graph of impulse train function is given in Fig. 1.5.
Continuous time periodic signals have Fourier series representation. Impulse train signal (function) also has Fourier series representation which can be written as

$$s(t) = \sum_{k=-\infty}^{\infty} S[k]e^{jk\frac{2\pi}{T_s}t} \qquad (1.7)$$

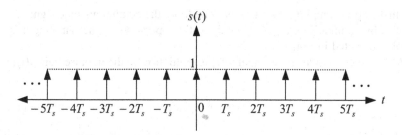

Fig. 1.5 Impulse train function

where $S[k]$ are the Fourier series coefficients which are calculated using

$$S[k] = \frac{1}{T_s} \sum_{k=-\infty}^{\infty} s(t)e^{-jk\frac{2\pi}{T_s}t}dt. \tag{1.8}$$

Let's now calculate the Fourier series coefficients of impulse train. Using (1.8) the Fourier series coefficients of the impulse train function can be calculated as

$$S[k] = \frac{1}{T_s} \int_{-\infty}^{\infty} \delta(t)e^{-jk\frac{2\pi}{T_s}t}dt \rightarrow S[k] = \frac{1}{T_s}e^0 \rightarrow S[k] = \frac{1}{T_s} \tag{1.9}$$

Replacing the calculated coefficients in (1.8) we get the Fourier series representation of the impulse train as

$$s(t) = \frac{1}{T_s} \sum_{k=-\infty}^{\infty} e^{jk\frac{2\pi}{T_s}t} \tag{1.10}$$

Using the Fourier series representation of the impulse train function, we can calculate its Fourier transform. For this purpose, we first need to know the Fourier transform of the exponential function. The Fourier transform of the exponential function is given as

$$e^{jw_0t} \overset{FT}{\longleftrightarrow} 2\pi\delta(w - w_0). \tag{1.11}$$

When the expression in (1.11) is used while taking the Fourier transform of (1.10), we obtain the Fourier transform of the impulse train

$$S(w) = \frac{2\pi}{T_s} \sum_{k=-\infty}^{\infty} \delta(w - kw_s), \quad w_s = \frac{2\pi}{T_s}. \tag{1.12}$$

1.2 Sampling Operation

The first step in sampling operation is to multiply the continuous time signal to be sampled by an impulse train. This multiplication operation for the sampling of sine signal is depicted in Fig. 1.6.

When the continuous time signal $x_c(t)$ is multiplied by the impulse train $s(t)$, we obtain

$$x_s(t) = x_c(t) \cdot s(t) \tag{1.13}$$

in which, if the explicit expression for the impulse train is inserted we get the mathematical expression

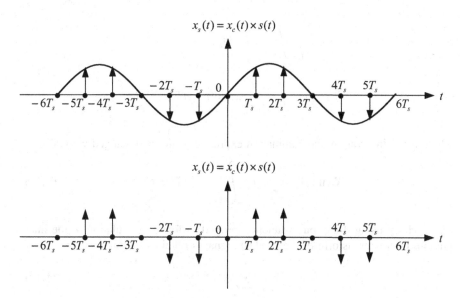

Fig. 1.6 Multiplication of sine signal by an impulse train

$$x_s(t) = x_c(t) \sum_{n=-\infty}^{\infty} \delta(t - nT_s) \tag{1.14}$$

which can be simplified using the impulse function property $\int f(t)\delta(t - t_0)dt = f(t_0)$ as

$$x_s(t) = \sum_{n=-\infty}^{\infty} x_c(nT_s)\delta(t - nT_s) \tag{1.15}$$

where substituting $x[n] = x_c(nT_s)$, we obtain

$$x_s(t) = \sum_{n=-\infty}^{\infty} x[n]\delta(t - nT_s) \tag{1.16}$$

1.2.1 The Fourier Transform of the Product Signal

We obtained the time domain expression for the product signal $x_s(t)$. Let's now consider the Fourier transform of the product signal $x_s(t)$. The Fourier transform of $x_s(t)$ is computed using

$$X_s(w) = \int_{-\infty}^{\infty} x_s(t)e^{-jwt}\,dt \rightarrow$$

$$X_s(w) = \int_{-\infty}^{\infty} \sum_{n=-\infty}^{\infty} x[n]\delta(t-nT_s)e^{-jwt}\,dt$$

(1.17)

where if the integration and summation expressions are interchanged we get

$$X_s(w) = \sum_{n=-\infty}^{\infty} x[n] \int_{-\infty}^{\infty} \delta(t-nT_s)e^{-jwt}\,dt \qquad (1.18)$$

on which by using the impulse function properties for the calculation of the integration, Fourier transform of the product signal is obtained as

$$X_s(w) = \sum_{n=-\infty}^{\infty} x[n]e^{-jwnT_s}. \qquad (1.19)$$

The right hand side of the (1.19) contains parameters from time domain.

However, there is not only one single expression for the Fourier transform of the product signal. We can find an alternative expression for the Fourier transform of product signal. Let's now find an alternative expression for the Fourier transform of product signal where both left and right sides only include expressions in frequency domain. Consider the product signal expression again

$$x_s(t) = x_c(t) \cdot s(t) \qquad (1.20)$$

where the right hand side is the product of two expressions, for this reason, the Fourier transform of $x_s(t)$ can be written as

$$X_s(w) = \frac{1}{2\pi}X_c(w) * S(w). \qquad (1.21)$$

where substituting the expression in (1.12) for $S(w)$, we get

$$X_s(w) = \frac{1}{2\pi}X_c(w) * \frac{2\pi}{T_s}\sum_{k=-\infty}^{\infty} \delta(w-kw_s) \qquad (1.22)$$

where by using the impulse function property and linearity of the convolution operation we obtain

$$X_s(w) = \frac{1}{T_s}\sum_{k=-\infty}^{\infty} X_c(w-kw_s). \qquad (1.23)$$

We have obtained a second alternative expression for the Fourier transform of product signal. Let's write both Fourier expressions again

$$X_s(w) = \sum_{n=-\infty}^{\infty} x[n]e^{-j2\pi nT_s} \quad X_s(w) = \frac{1}{T_s} \sum_{k=-\infty}^{\infty} X_c(w - kw_s). \tag{1.24}$$

In these expressions the left hand sides are both $X_s(w)$. So the right hand sides should also be equal to each other. Equating the right hand sides of the expressions in (1.24), we obtain the equation

$$\sum_{n=-\infty}^{\infty} x[n]e^{-jwnT_s} = \frac{1}{T_s} \sum_{k=-\infty}^{\infty} X_c(w - kw_s). \tag{1.25}$$

The Fourier transform of the digital signal $x[n]$ is calculated using

$$X_n(w) = \sum_{n=-\infty}^{\infty} x[n]e^{-jwn}$$

which resembles to the left term in (1.25). We can write the left hand side of (1.25) in terms of $X_n(w)$ as

$$\underbrace{\sum_{n=-\infty}^{\infty} x[n]e^{-jwnT_s}}_{X_n(wT_s)} = \frac{1}{T_s} \sum_{k=-\infty}^{\infty} X_c(w - kw_s) \tag{1.26}$$

which yields

$$X_n(wT_s) = \frac{1}{T_s} \sum_{k=-\infty}^{\infty} X_c(w - kw_s) \tag{1.27}$$

from which $X_n(w)$ can be obtained by replacing w with $\frac{w}{T_s}$ and we obtain

$$X_n(w) = \frac{1}{T_s} \sum_{k=-\infty}^{\infty} X_c\left(\frac{w}{T_s} - kw_s\right). \tag{1.28}$$

In the expression (1.28) the left hand side represents the Fourier transform of the digital signal obtained from an analog signal via sampling operation. In other words, it represents the Fourier transform of the mathematical sequence obtained from analog signal via sampling operation. The right hand side consists of shifted and scaled replicas of $X_c(w)$ which is the Fourier transform of analog signal on which sampling operation is performed. Since $X_n(w)$ is the Fourier transform of a digital signal, it is periodic with period 2π. If the digital signal is also periodic in time domain, then its Fourier transform is periodic with period 2π consisting of impulses spaced by multiples of 2π.

Now, let's summarize the formulas we have derived up to this point.

In Time Domain

Continuous time signal $x_c(t)$

Sampling operation $x[n] = x_c(nT_s)$

Sampling period T_s

Impulse train $s(t) = \sum_{n=-\infty}^{\infty} \delta(t - nT_s)$

Product signal $x_s(t) = x_c(t) \cdot s(t)$

Product signal $x_s(t) = \sum_{k=-\infty}^{\infty} x_c(nT_s)\delta(t - nT_s)$

Product signal $x_s(t) = \sum_{k=-\infty}^{\infty} x[n]\delta(t - nT_s)$

In Frequency Domain

Fourier transform of product function $x_s(t)X_s(w) = \int_{-\infty}^{\infty} x_s(t)e^{-jwt}dt$

Fourier transform of product function $x_s(t)X_s(w) = \frac{1}{2\pi}X_c(w) * S(w)$

Sampling frequency in rad/sec $w_s = \frac{2\pi}{T_s}$

Fourier transform of product function $x_s(t)X_s(w) = \frac{1}{T_s}\sum_{k=-\infty}^{\infty} X_c(w - kw_s)$

Fourier transform of $x[n]$ digital signal $X_n(w) = \sum_{n=-\infty}^{\infty} x[n]e^{-jwn}$

Fourier transform of $x[n]$ digital signal $X_n(w) = \frac{1}{T_s}\sum_{k=-\infty}^{\infty} X_c\left(\frac{w}{T_s} - kw_s\right)$

Exercise: Given the digital signal $x[n] = \begin{bmatrix} 2 & \underset{n=0}{3.5} & -4 & 5 & 6 & 3 & -2 \end{bmatrix}$

draw the graphs of

(a) $y(t) = \sum_{n=-\infty}^{\infty} x[n]\delta(t - nT_s)$ where $T_s = 1/4$ s.

(b) $g(t) = \sum_{n=-\infty}^{\infty} x[2n]\delta(t - nT_s)$ where $T_s = 1/8$ s.

(c) $h(t) = \sum_{n=-\infty}^{\infty} x[n/2]\delta(t - nT_s)$ where $T_s = 1/4$ s.

Exercise: Calculate the Fourier transforms of

$$x_c(t) = \delta(t) + \delta(t - 1)$$

and

$$x[n] = \delta[n] + \delta[n - 1]$$

and draw their magnitude and phase responses.

1.3   How to Draw Fourier Transforms of Product Signal and Digital Signal

The derived mathematical expression for $X_s(w)$ is given as

$$X_s(w) = \frac{1}{T_s} \sum_{k=-\infty}^{\infty} X_c(w - kw_s) \qquad (1.29)$$

which is a periodic function and one period of this function around the origin, assuming no overlapping among shifted replicas, can be written as

$$\frac{1}{T_s} X_c(w). \qquad (1.30)$$

The period of $X_s(w)$ is denoted by w_s whose value equals to $\frac{2\pi}{T_s}$. Drawing the graph of $X_s(w)$ consists of two steps. In the first step, we draw the graph of $\frac{1}{T_s} X_c(w)$ around the origin. Then in the next step, the drawn graph around the origin is shifted to the left and right by integer multiples of $w_s = \frac{2\pi}{T_s}$, i.e., by $kw_s, k \in Z$, and the shifted replicas together with the one around the origin are all summed.

Before studying some problems on the drawing of $X_s(w)$, let's inspect some examples to prepare ourselves for the drawing of $X_s(w)$.

Example 1.3 In Fig. 1.7, the graphics of $X_1(w)$ and $X_2(w)$ are given for the interval $0 \le w \le 4$. Draw the graph of $X_1(w) + X_2(w)$ for the same interval.

Solution 1.3 To draw the graphic of $X_1(w) + X_2(w)$, let's first write the mathematical expressions for each function, then sum these functions to get the mathematical expression for the summed signals. The mathematical expressions for the signals $X_1(w)$ and $X_2(w)$ are given as

$$X_1(w) = -\frac{w}{2} + 2 \quad X_2(w) = \frac{w}{2}$$

If we sum mathematical expressions for the signals $X_1(w)$ and $X_2(w)$, we get

$$X_1(w) + X_2(w) = 2.$$

Fig. 1.7 The graphics of $X_1(w)$ and $X_2(w)$

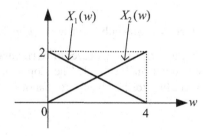

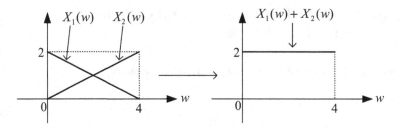

Fig. 1.8 The graphics of $X_1(w) + X_2(w)$

Fig. 1.9 The graphics of $X_1(w)$ and $X_2(w)$ functions

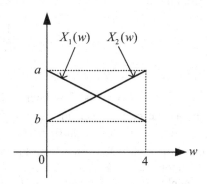

The obtained result is graphically shown in Fig. 1.8.

Example 1.4 The graphics of $X_1(w)$ and $X_2(w)$ for the interval $0 \le w \le 4$ are shown in Fig. 1.9. The slopes of the lines in Fig. 1.9 are $-1/2$ and $1/2$. According to the given information, draw the graphic of $X_1(w) + X_2(w)$ for the same interval.

Solution 1.4 To draw the graph of $X_1(w) + X_2(w)$ we need to find its mathematical expression. For this purpose, let's first write the mathematical expressions for $X_1(w)$ and $X_2(w)$ using the given information for the interval $0 \le w \le 4$ as

$$X_1(w) = -\frac{w}{2} + a \quad X_2(w) = \frac{w}{2} + b$$

When the mathematical expressions for $X_1(w)$ and $X_2(w)$ are summed, we obtain

$$X_1(w) + X_2(w) = a + b$$

which is graphically depicted in Fig. 1.10.

Example 1.5 The graphics of $X_1(w)$ and $X_2(w)$ functions for the interval $0 \le w \le 4$ are shown in Fig. 1.11. The slopes of the lines in Fig. 1.11 are $-1/2$ and $1/2$. According to the given information, draw the graphic of $X_1(w) + X_2(w)$.

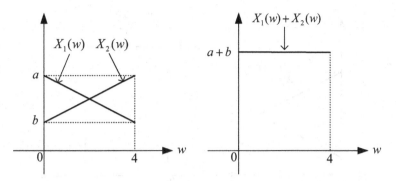

Fig. 1.10 The graphic of $X_1(w) + X_2(w)$

Fig. 1.11 The graphic of $X_1(w)$ and $X_2(w)$

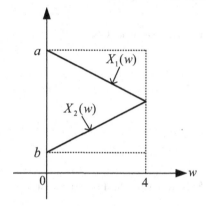

Solution 1.5 We can follow the same steps as in the previous two examples. The line equations of $X_1(w)$ and $X_2(w)$ can be written as

$$X_1(w) = -mw + a \quad X_2(w) = mw + b.$$

If we sum the line equations of these two functions, we obtain

$$X_1(w) + X_2(w) = a + b.$$

The obtained result is depicted in Fig. 1.12. We will use this result to draw the graphs of the digital signals having the spectral overlapping problem.

Example 1.6 $x_c(t)$ is a continuous time signal and its Fourier transform is denoted by $X_c(w)$. The graph of $X_c(w)$ is depicted in Fig. 1.13. As it is seen from the Fourier transform graph, $x_c(t)$ is a low-pass signal with bandwidth w_N.

Let $x_s(t) = x_c(t) \times s(t)$ where $s(t)$ is the impulse train signal. Draw the Fourier transform of $x_s(t)$ assuming that $w_s > 2w_N$, i.e., draw $X_s(w)$.

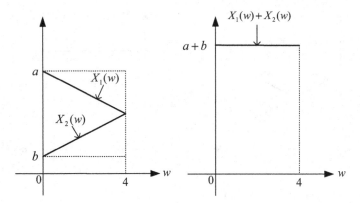

Fig. 1.12 The graphic of $X_1(w) + X_2(w)$

Fig. 1.13 Graph of $X_c(w)$

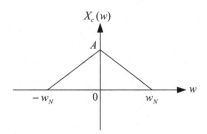

Solution 1.6 The Fourier transform of the product signal $x_s(t)$ is

$$X_s(w) = \frac{1}{T_s} \sum_{k=-\infty}^{\infty} X_c(w - kw_s)$$

which is a periodic function with period $w_s = \frac{2\pi}{T_s}$. When the summation expression in $X_s(w)$ is expanded, we get

$$X_s(w) = \cdots + \frac{1}{T_s} X_c(w + w_s) + \frac{1}{T_s} X_c(w) + \frac{1}{T_s} X_c(w - w_s) + \cdots$$

where the graphs of the terms $\frac{1}{T_s} X_c(w), \frac{1}{T_s} X_c(w + w_s)$, and $\frac{1}{T_s} X_c(w - w_s)$ are depicted in Fig. 1.14.

The other shifted and scaled replicas can be drawn in a similar manner as in Fig. 1.14. When the shifted and scaled replicas are summed, we obtain the graphic of $X_s(w)$ as depicted in Fig. 1.15.

Example 1.7 $x_c(t)$ is a continuous time signal and its Fourier transform $X_c(w)$ is depicted in Fig. 1.16. $x_c(t)$ is sampled by the sampling period $T_s = \frac{1}{2000}$ s. Draw the graph of $X_s(w)$

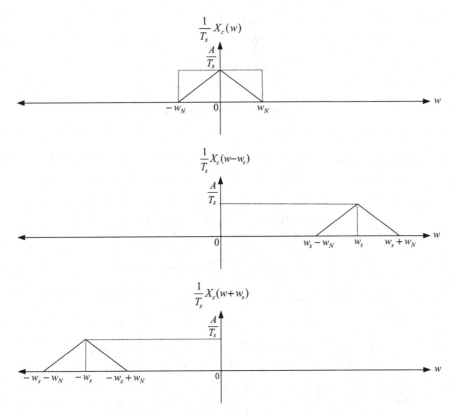

Fig. 1.14 The graphics of $\frac{1}{T_s}X_c(w), \frac{1}{T_s}X_c(w+w_s)$ and $\frac{1}{T_s}X_c(w-w_s)$

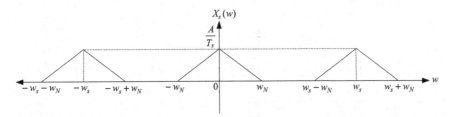

Fig. 1.15 The graphic of $X_s(w)$

Solution 1.7 The sampling frequency in rad/sec is

$$w_s = \frac{2\pi}{T_s} \rightarrow w_s = 4000\pi \, \text{rad/s}$$

The shifted $X_c(w)$ signals by multiples of w_s are shown in Fig. 1.17.

As it is clear from Fig. 1.17, shifted replicas overlap. Summing the overlapped amplitudes, we obtain the signal shown in Fig. 1.18.

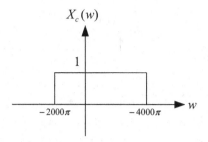

Fig. 1.16 The graphic of $X_c(w)$

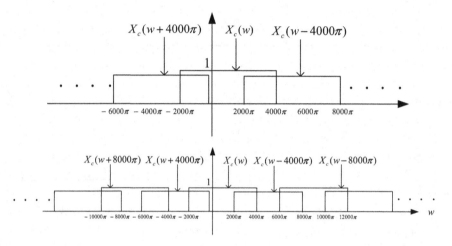

Fig. 1.17 Shifted $X_c(w)$ signals

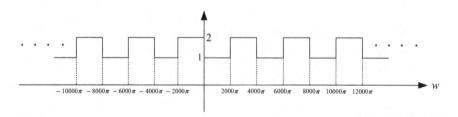

Fig. 1.18 Summation of the shifted replicas

In the last stage, we divide the amplitudes of the summed signal shown in Fig. 1.18 by T_s. Since $T_s = \frac{1}{2000}$ dividing the amplitudes by T_s equals to multiplying the amplitudes by 2000. After multiplying the amplitudes by 2000 we obtain the graphic of the function $X_s(w)$ as depicted in Fig. 1.19.

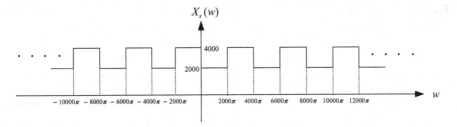

Fig. 1.19 The graphic of $X_s(w)$

Fig. 1.20 The graphic
of $X_c(w)$

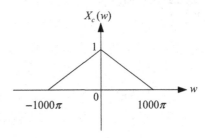

Example 1.8 The graphic of $X_c(w)$ is shown in Fig. 1.20. Draw the graphic of

$$X_s(w) = 250 \sum_{k=-\infty}^{\infty} X_c(w - k500\pi)$$

Solution 1.8 From the equation

$$X_s(w) = 250 \sum_{k=-\infty}^{\infty} X_c(w - k500\pi)$$

it is seen that the sampling frequency in rad/sec is $w_s = 500\pi$ rad/s. Let's partition the horizontal axis of $X_c(w)$ as in Fig. 1.21 considering the sampling frequency value.

Now let's draw the shifted $X_c(w)$ signals as shown in Fig. 1.22.

The graphs of $X_c(w)$, $X_c(w - w_s)$ and $X_c(w + w_s)$ altogether are given in Fig. 1.23.

More shifted graphs of $X_c(w)$ are given in Fig. 1.24.

If the above graph is carefully inspected, it is seen that a portion of the graph repeats itself along the horizontal axis. The repeated part is indicated by bold lines in Fig. 1.25.

Now let's write the mathematical equations for the line segments, a, b, c, d, e, f, g, h appearing in the repeating pattern in Fig. 1.25 as

Fig. 1.21 The graphic
of $X_c(w)$

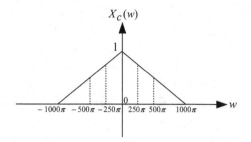

Fig. 1.22 Shifted graphs
of $X_c(w)$

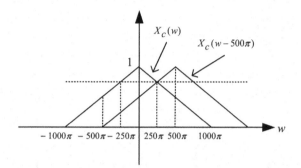

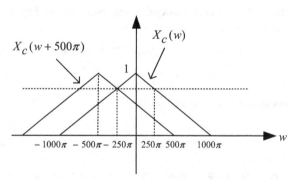

Fig. 1.23 Shifted graphs
of $X_c(w)$

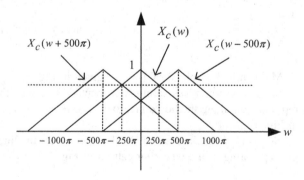

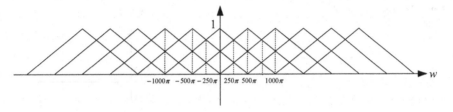

Fig. 1.24 Shifted graphs of $X_c(w)$

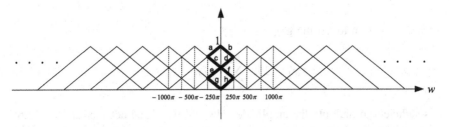

Fig. 1.25 Shifted graphs of $X_c(w)$ and repeating pattern

$$m = \frac{1}{1000\pi} \quad -250\pi \leq w \leq 250\pi$$

$$y_a = mw + 1 \quad y_b = -mw + 1 \quad y_c = -mw + \frac{1}{2} \quad y_d = mw + \frac{1}{2}$$

$$y_e = mw + \frac{1}{2} \quad y_f = -mw + \frac{1}{2} \quad y_g = -mw \quad y_h = mw.$$

If we sum the equations for the line segments, a, c, e, g and b, d, f, g we get the results

$$y_a + y_c + y_e + y_g = 2 \quad y_b + y_d + y_f + y_g = 2.$$

and the graph of $X_s(w)$ is drawn as in Fig. 1.26.

Fig. 1.26 Summation result of shifted $X_c(w)$ functions

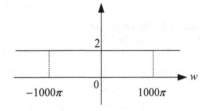

Fig. 1.27 The graphic of
$X_s(w)$

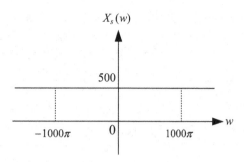

In the last step to get the graph of

$$X_s(w) = 250 \sum_{k=-\infty}^{\infty} X_c(w - k500\pi)$$

it is sufficient to multiply the amplitude values of the signal depicted in Fig. 1.26. After amplitude multiplication, we obtain the graph of $X_s(w)$ as depicted in Fig. 1.27.

Example 1.9 The graphic of $X_c(w)$ is shown in Fig. 1.28. Draw the graphic of

$$X_s(w) = \frac{1}{T_s} \sum_{k=-\infty}^{\infty} X_c\left(w - k\frac{2\pi}{T_s}\right)$$

for $T_s = \frac{1}{375}$ s.

Solution 1.9 We can write the sampling frequency in rad/san unit as $w_s = \frac{2\pi}{T_s} \rightarrow w_s = 750\pi \text{rad/s}$. In the next step, we shift the function $X_c(w)$ to the left and right by $kw_s, k \in Z$. Some shifted replicas of $X_c(w)$ are displayed in Fig. 1.29.

If the graph in Fig. 1.29 is inspected carefully, it can be seen that a define pattern repeats itself along the shape. The repeating pattern is indicated in bold lines in Fig. 1.30.

The repeating pattern in Fig. 1.30 is redrawn alone in Fig. 1.31 in details.

Fig. 1.28 The graphic of
$X_c(w)$ for Example 1.9

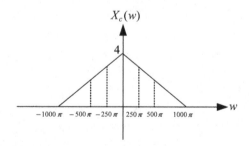

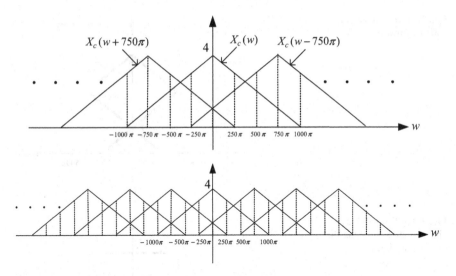

Fig. 1.29 Shifted $X_c(w)$ functions

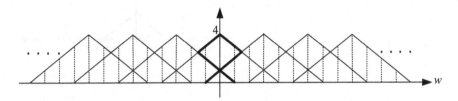

Fig. 1.30 The repeating pattern

If the graphic in Fig. 1.31 is inspected carefully, it is seen that the line pairs in the upper left and upper right shadowed rectangles overlap each other and their slopes are equal in magnitude but opposite in sign. For this reason, the sum of the line equations for line pairs is a constant number and it equals to $1 + 4 = 5$. After summing the overlapping line equations, we get the graphic in Fig. 1.32.

If the triangle shape and horizontal line in Fig. 1.32 are summed, we get the graphic in Fig. 1.33.

The graphic shown in Fig. 1.33 corresponds to one period of the function $X_s(w)$ around origin. If one period of $X_s(w)$ around origin is shifted to the right and left by multiples of $w_s = 750\pi$ and shifted replicas are all summed together with the graph around origin, we get the graphic of $X_s(w)$ as in Fig. 1.34.

Solution 2 In fact, the second solution provided here is more complex than the first solution. However, we find it useful to illustrate the different perspectives for the solution of a problem.

The repeating pattern chosen in solution can be interpreted in a different manner. In fact, the interpretation of the repeating patterns depends on the reader's

Fig. 1.31 The repeating
pattern drawn in details

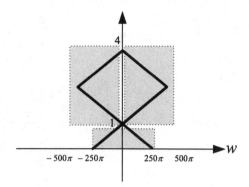

Fig. 1.32 The graphic
obtained after summing the
overlapping lines

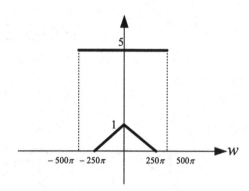

Fig. 1.33 The graphic
obtained after summing the
triangle shape and horizontal
line

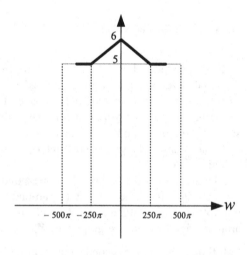

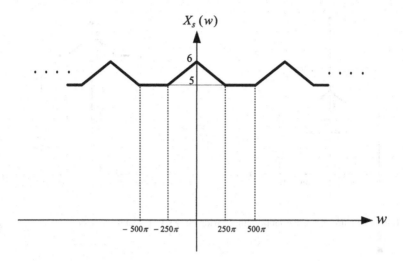

Fig. 1.34 The graphic of $X_s(w)$.

Fig. 1.35 Repeating part in details

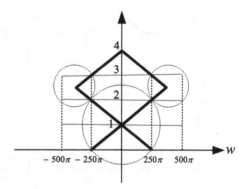

perception. The overlapped lines in the repeating pattern are shown inside circles in Fig. 1.35 in a different approach than the one in solution 1.

In Fig. 1.35 the sum of the overlapped lines inside circles results in constant numbers, and when the constants are added to the top triangle shape, we obtain one period of $X_s(w)$ around origin. This is illustrated in Fig. 1.36.

When the obtained one period around the origin is shifted to the left and right, we obtain $X_s(w)$ function in Fig. 1.37.

Exercise: The graphic of $X_c(w)$ function is depicted in Fig. 1.38. Using the given figure draw the graph of

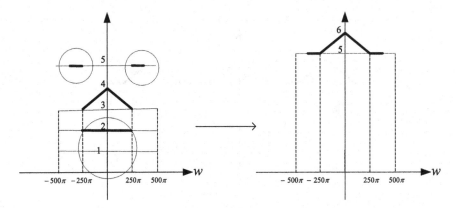

Fig. 1.36 The sum of the overlapped lines inside circles in repeating pattern

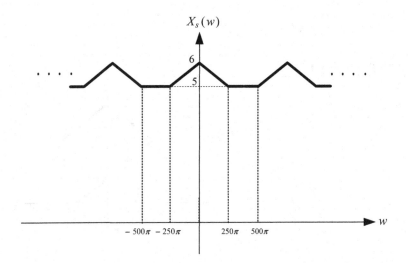

Fig. 1.37 $X_s(w)$ graph

Fig. 1.38 Fourier transform
of an input signal

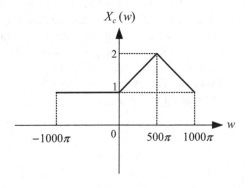

$$X_s(w) = 500 \sum_{k=-\infty}^{\infty} X_c(w - k1000\pi).$$

Exercise: The Fourier transform of a continuous time signal is given as

$$X_c(w) = \pi(\delta(w - 500\pi) + \delta(+500\pi)).$$

Using the given Fourier transform draw the graph of

$$X_s(w) = 200 \sum_{k=-\infty}^{\infty} X_c(w - k400\pi).$$

1.3.1 Drawing the Fourier Transform of Digital Signal

Assume that $X_n(w)$ is the Fourier transform of $x[n]$ which is obtained from $x_c(t)$ via sampling operation, i.e., $x[n] = x_c(t)|_{t=nT_s} \rightarrow x[n] = x_c(nT_s)$ and the mathematical expression for $X_n(w)$ is given as

$$X_n(w) = \frac{1}{T_s} \sum_{k=-\infty}^{\infty} X_c\left(\frac{w}{T_s} - kw_s\right). \tag{1.31}$$

To draw the graph of $X_n(w)$ two different methods can be followed. Below, we explain these two methods separately.

Method 1: First draw the graph of $X_s(w)$, i.e., draw the Fourier transform of the product signal $x_s(t) = x_c(t)s(t)$ as discussed in the previous section. Once you have the graph of $X_s(w)$, to get the graph of $X_n(w)$, multiply the horizontal axis of $X_s(w)$ by sampling period T_s.

Method 2: Since $X_n(w)$ is the Fourier transform of the digital signal $x[n]$, it is a periodic signal and its period equals 2π. To draw the graph of $X_n(w)$, first draw the graph of $\frac{1}{T_s}X_c\left(\frac{w}{T_s}\right)$ around origin, then shift the drawn signal to the left and right by multiples of 2π, and sum the shifted replicas. Note that to draw the graph of $\frac{1}{T_s}X_c\left(\frac{w}{T_s}\right)$, we multiply the amplitude values of $X_c(w)$ by $1/T_s$ and multiply horizontal axis of $X_c(w)$ by T_s, i.e., divide the horizontal axis of $X_c(w)$ by $1/T_s$.

Let's now provide some examples to comprehend the subject better.

Example 1.10 The Fourier transform of a continuous time signal $x_c(t)$ is depicted in Fig. 1.39. Draw $X_n(w)$, the Fourier transform of $x[n] = x_c(nT_s)$ where T_s is the sampling period. Assume that $w_s > 2w_N$.

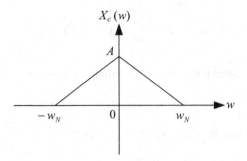

Fig. 1.39 Fourier transform of a low pass input signal

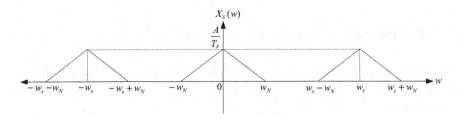

Fig. 1.40 Graph of $X_s(w)$

Solution 1.10
Method 1: Let's first draw the graph of

$$X_s(w) = \frac{1}{T_s} \sum_{k=-\infty}^{\infty} X_c(w - kw_s)$$

which is a periodic function with period $w_s = \frac{2\pi}{T_s}$. The graph of $X_s(w)$ is shown in Fig. 1.40.

In the second step, we multiply the horizontal axis of $X_s(w)$ by T_s to get the graph of $X_n(w)$. The graph of $X_n(w)$ is shown in Fig. 1.41.

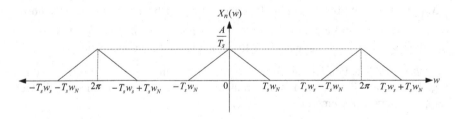

Fig. 1.41 Graph of $X_n(w)$

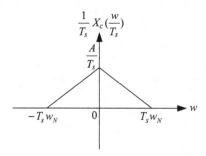

Fig. 1.42 The graph of $\frac{1}{T_s} X_c\left(\frac{w}{T_s}\right)$

Method 2: In the second method, we first draw the graph of

$$\frac{1}{T_s} X_c\left(\frac{w}{T_s}\right)$$

then we shift the drawn graph to the left and right by multiples of 2π and obtain the graph of $X_n(w)$. To draw the graph of $\frac{1}{T_s} X_c\left(\frac{w}{T_s}\right)$, we multiply the vertical and horizontal axes of $X_c(w)$ by $\frac{1}{T_s}$ and T_s respectively. In Fig. 1.42 the graph of $\frac{1}{T_s} X_c\left(\frac{w}{T_s}\right)$ is depicted.

Let's denote $\frac{1}{T_s} X_c\left(\frac{w}{T_s}\right)$ by $X_{n1}(w)$. To get the graph of $X_n(w)$, we shift $X_{n1}(w)$ to the left and right by multiples of 2π and sum the shifted replicas. This operation is illustrated in Fig. 1.43.

Example 1.11 The continuous time signal $x_c(t)$ is given as

$$x_c(t) = \cos(4000\pi t).$$

(a) Draw $X_c(w)$, the Fourier transform of $x_c(t)$.
(b) Let $x_s(t) = x_c(t)s(t)$ where $s(t)$ is the impulse train and $T_s = \frac{1}{8000}$ s. Draw $X_s(w)$, the Fourier transform of $x_s(t)$.
(c) Let $x[n] = x_c(nT_s)$ where $T_s = \frac{1}{8000}$ s. Draw $X_n(w)$, the Fourier transform of $x[n]$.

Solution 1.11 Before computing the Fourier transform of the given cosine signal, let's review some properties of the exponential signal. The Fourier transform of an exponential signal is given as

$$e^{jw_N t} \xrightarrow{FT} 2\pi\delta(w - w_N) \tag{1.32}$$

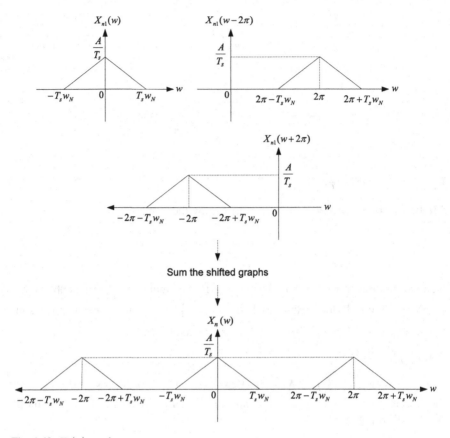

Fig. 1.43 $X_n(w)$ graph

and sine and cosine signals can be written in terms of the exponential signals as

$$\sin(w_N t) = \frac{1}{2j}\left(e^{+jw_N t} - e^{-jw_N t}\right) \quad \cos(w_N t) = \frac{1}{2}\left(e^{+jw_N t} + e^{-jw_N t}\right) \quad (1.33)$$

And the Fourier transforms of the sinusoidal signals are given as

$$\sin(w_N t) \xrightarrow{FT} \frac{\pi}{j}\left(\delta(w - w_N) - \delta(w + w_N)\right)$$

$$\cos(w_N t) \xrightarrow{FT} \pi(\delta(w - w_N) + \delta(w + w_N)). \quad (1.34)$$

(a) Since we refreshed some background information we can start to solve our problem. The Fourier transform of $x_c(t) = \cos(4000\pi t)$ can be calculated as

$$\cos(4000\pi t) \xrightarrow{FT} \pi(\delta(w - 4000\pi) + \delta(w + 4000\pi))$$

and its graph is depicted as in Fig. 1.44.

(b) Since $x_s(t) = x_c(t)s(t)$, and $s(t) = \sum_{k=-\infty}^{\infty} \delta(t - kT_s)$, where T_s is the sampling period, Fourier transform of $x_s(t)$ is

$$X_s(w) = \frac{1}{T_s} \sum_{k=-\infty}^{\infty} X_c(w - kw_s)$$

where $w_s = \frac{2\pi}{T_s} \rightarrow \frac{2\pi}{1/8000} = 16000\pi$. Using the Fourier transform expression $X_c(w)$ found in the previous part, $X_s(w)$ can be calculated as

$$X_s(w) = 8000 \sum_{k=-\infty}^{\infty} X_c(w - k16000\pi) \rightarrow$$

$$X_s(w) = 8000\pi \sum_{k=-\infty}^{\infty} (\delta(w - 4000\pi - k16000\pi) + \delta(w + 4000\pi - k16000\pi))$$

and the graph of $X_s(w)$ is displayed in Fig. 1.45.

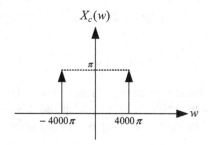

Fig. 1.44 Fourier transform of $x_c(t) = \cos(4000\pi t)$

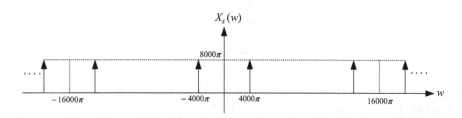

Fig. 1.45 Fourier transform of the product signal $x_s(t)$

Fig. 1.46 Fourier transform
of $X_n(w)$

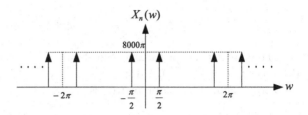

(c) To get the graph of $X_n(w)$, it is sufficient to multiply the horizontal axis of
$X_s(w)$ by T_s. Thus, the graph of $X_n(w)$ is obtained as in Fig. 1.46.

1.4 Aliasing (Spectral Overlapping)

Let the Fourier transform of a continuous time signal be as given as in Fig. 1.47.
Using the Fourier transform in Fig. 1.47, let's draw the graph of

$$X_s(w) = \frac{1}{T_s} \sum_{k=-\infty}^{\infty} X_c(w - kw_s) \tag{1.35}$$

as in Fig. 1.48.

It is clear from Fig. 1.48 that the condition for the shifted graphs not to overlap
can be written as

$$w_s - w_1 > w_2 \rightarrow w_s > w_1 + w_2 \tag{1.36}$$

and if $w_1 < w_2$ then no aliasing condition in (1.36) can also be written as $w_s > 2w_2$.

If $w_s < w_1 + w_2$, then the shifted graphs overlap and this condition is named as
aliasing (overlapping). The case of aliasing is depicted in Fig. 1.49.

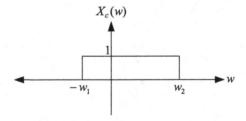

Fig. 1.47 The Fourier transform of a low pass signal

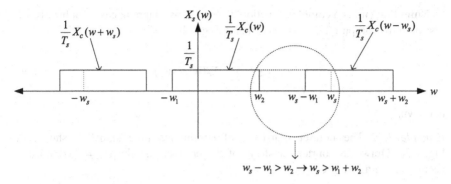

Fig. 1.48 The graph of $X_s(w)$

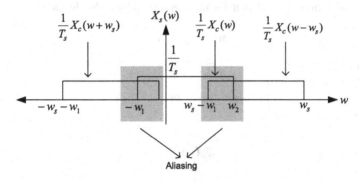

Fig. 1.49 Aliasing case

For many signals the Fourier transform is symmetric with respect to the vertical axis, i.e., $w_1 = w_2$. And for the symmetric case, let $w_1 = w_2 = w_N$ and the condition for no aliasing in this case can be stated as

$$w_s > 2w_N \tag{1.37}$$

where the unit of the frequencies is rad/sec. If we write the explicit expressions for the frequencies in (1.37), we get

$$\frac{2\pi}{T_s} > 2\frac{2\pi}{T_N} \tag{1.38}$$

and the condition for no aliasing can be written as $f_s > 2f_N$. This means that for no aliasing, the sampling frequency in unit of Hertz should be greater than twice of the highest frequency available in the signal.

Note: If $X_c(w)$ is a complex function, to see the overlapping case graphically we first draw the graph of $|X_c(w)|$, and then the graph of

$$X_s(w) = \frac{1}{T_s} \sum_{k=-\infty}^{\infty} |X_c(w - kw_s)| \tag{1.39}$$

is drawn.

Example 1.12 The Fourier transform of continuous time signal is shown in Fig. 1.50. Draw the Fourier transform of the product signal $x_s(t) = x_c(t)s(t)$ and decide on the aliasing case.

Solution 1.12 The graph of $X_s(w) = \frac{1}{T_s}\sum_{k=-\infty}^{\infty} X_c(w - kw_s)$ is depicted in Fig. 1.51.

It is clear from Fig. 1.51 that for no overlapping, we should have

$$w_s - w_N > w_N \tag{1.40}$$

leading to

$$w_s > 2w_N \tag{1.41}$$

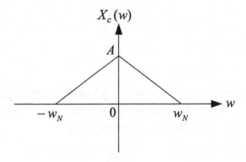

Fig. 1.50 Graph of $X_c(w)$

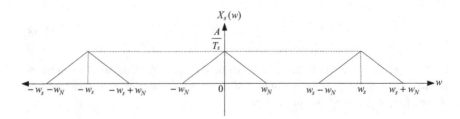

Fig. 1.51 Graph of $X_s(w)$

and no aliasing condition in (1.41) can also be expressed as

$$w_s > 2w_N \rightarrow 2\pi f_s > 2 \times 2\pi f_N \rightarrow f_s > 2f_N. \tag{1.42}$$

1.4.1   The Meaning of the Aliasing (Overlapping)

Sampling frequency implies the number of samples taken per-second from a continuous time signal. The collected samples are either transmitted, stored, or processed, and the analog signal can be reconstructed from the digital samples.

If sampling frequency is not high enough, the analog signal cannot be reconstructed due to insufficient number of received samples or it can only be partially reconstructed. In frequency domain, the effect of insufficient number of samples is seen as aliasing or spectral overlapping.

Example 1.13 The continuous time signal $x_c(t) = cos(20\pi t) + sin(40\pi t)$ is to be sampled. Choose a sampling frequency such that no aliasing occurs for the generated digital signal in frequency domain.

Solution 1.13 Let's first calculate the Fourier transform of the continuous time signal. For this purpose, the Fourier transforms of sinusoidal signals are reminded as

$$cos(w_0 t) \overset{FT}{\leftrightarrow} \pi(\delta(w - w_0) + \delta(w + w_0))$$
$$sin(w_0 t) \overset{FT}{\leftrightarrow} \frac{\pi}{j}(\delta(w - w_0) - \delta(w + w_0))$$

where substituting $w_0 = 2\pi f_0, w = 2\pi f$, we get the alternative form for the Fourier transform of the sinusoidal signals as

$$cos(2\pi f_0 t) \overset{FT}{\leftrightarrow} \frac{1}{2}(\delta(f - f_0) + \delta(f + f_0))$$
$$sin(2\pi f_0 t) \overset{FT}{\leftrightarrow} \frac{1}{2j}(\delta(f - f_0) - \delta(f + f_0))$$

While obtaining the alternative forms, we made use of the property

$$\delta(2\pi(f - f_0)) = \frac{1}{2\pi}\delta(f - f_0). \tag{1.43}$$

Using the Fourier transform formulas for the sinusoidal signals, we can calculate the Fourier transform of the continuous time signal given in the example and plot its graph as in Fig. 1.52.

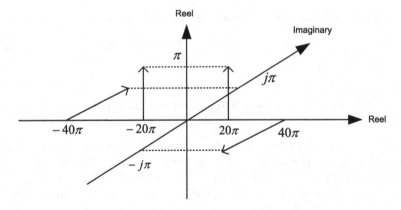

Fig. 1.52 Fourier transform of the composite signal $x_c(t)$

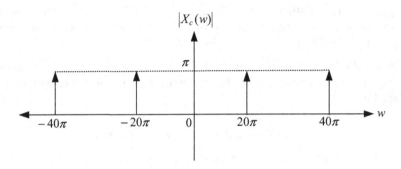

Fig. 1.53 Graph of $|X_c(w)|$

The Fourier transform of the summed sinusoids given in Fig. 1.52 seems to be complex to judge although not impossible. For easiness of the illustration, let's take the absolute value of the Fourier transforms and depict them as in Fig. 1.53.

As it is seen from Fig. 1.53 that the highest frequency available in the continuous time signal $x_c(t)$ is 40π rad/s or 20 Hz and the lowest positive frequency is 0. The analog signal is a low pass signal. The sampling frequency preventing aliasing should satisfy

$$w_s > 2 \times 40\pi$$

or in terms of unit of Hz, $f_s > 40$ Hz.

Method 2: Comparing the given sinusoidal functions to $\cos(2\pi f_1 t)$ and $\sin(2\pi f_2 t)$ expressions, we find the frequencies of the sinusoidal signals as $f_1 = 10$ Hz and $f_2 = 20$ Hz, and decide on the sampling frequency as

$$f_s > 2 \times 20 \text{ Hz} \to f_s > 40 \text{ Hz}$$

Example 1.14 If $x[n] = x_c(nT_s)$ then the Fourier transform of $x[n]$ is written as

$$X_n(w) = \frac{1}{T_s} \sum_{k=-\infty}^{\infty} X_c\left(\frac{w}{T_s} - kw_s\right) \tag{1.44}$$

where $X_c(w)$ is the Fourier transform of continuous time signal $x_c(t)$. The Fourier transform of the digital signal $x[n]$ can also be calculated using the Fourier transform formula directly, i.e.,

$$X_n(w) = \sum_{n=-\infty}^{\infty} x[n]e^{-jwn} \tag{1.45}$$

Derive (1.44) starting from the right hand side of (1.45).

Solution 1.14 Before starting to the derivation, let's remember the Fourier and inverse Fourier transforms of continuous time signal

$$X_c(w) = \int_{-\infty}^{\infty} x_c(t)e^{-jwt}dt \quad x_c(t) = \frac{1}{2\pi} \int_{-\infty}^{\infty} X_c(w)e^{jwt}dw.$$

If the time parameter 't' is replaced by 'nT_s' in inverse Fourier transform expression, we get

$$x_c(nT_s) = \frac{1}{2\pi} \int_{-\infty}^{\infty} X_c(w)e^{jwnT_s}dw. \tag{1.46}$$

For the digital signal $x[n]$, we have the Fourier transform expression

$$X_n(w) = \sum_{n=-\infty}^{\infty} x[n]e^{-jwn} \tag{1.47}$$

in which if we substitute $x[n] = X_c(nT_s)$, we get

$$X_n(w) = \sum_{n=-\infty}^{\infty} x_c(nT_s)e^{-jwn}. \tag{1.48}$$

In (1.48) if $x_c(nT_s)$ is replaced by (1.46), we get

$$X_n(w) = \sum_{n=-\infty}^{\infty} \frac{1}{2\pi} \int_{-\infty}^{\infty} X_c(\lambda) e^{j\lambda n T_s} d\lambda \, e^{-jwn} \tag{1.49}$$

which can be re-arranged as

$$X_n(w) = \frac{1}{2\pi} \sum_{n=-\infty}^{\infty} X_c(\lambda) \int_{-\infty}^{\infty} e^{-j(w-\lambda T_s)n} d\lambda \tag{1.50}$$

and exchanging the places of summation and integration operators, we obtain

$$X_n(w) = \frac{1}{2\pi} \int_{-\infty}^{\infty} X_c(\lambda) \sum_{n=-\infty}^{\infty} e^{-j(w-\lambda T_s)n} d\lambda \tag{1.51}$$

on which we can use the property

$$\sum_{n=-\infty}^{\infty} e^{-j(w-\lambda T_s)n} = 2\pi \sum_{k=-\infty}^{\infty} \delta(w - \lambda T_s - k2\pi) \tag{1.52}$$

$$\delta(w - \lambda T_s - k2\pi) = \delta\left(T_s\left(\frac{w}{T_s} - \lambda - k\frac{2\pi}{T_s} \right) \right)$$
$$= \frac{1}{T_s}\delta\left(\frac{w}{T_s} - \lambda - k\frac{2\pi}{T_s} \right) \tag{1.53}$$

$$\sum_{n=-\infty}^{\infty} e^{-j(w-\lambda T_s)n} = \frac{2\pi}{T_s} \sum_{k=-\infty}^{\infty} \delta\left(\frac{w}{T_s} - \lambda - k\frac{2\pi}{T_s} \right) \tag{1.54}$$

leading to the expression

$$X_n(w) = \frac{1}{2\pi} \int_{-\infty}^{\infty} X_c(\lambda) \frac{2\pi}{T_s} \sum_{k=-\infty}^{\infty} \delta\left(\frac{w}{T_s} - \lambda - k\frac{2\pi}{T_s} \right) d\lambda \tag{1.55}$$

where upon exchanging summation and integration operators, we get

$$X_n(w) = \frac{1}{T_s} \sum_{k=-\infty}^{\infty} \frac{1}{2\pi} \int_{-\infty}^{\infty} X_c(\lambda)\delta\left(\frac{w}{T_s} - \lambda - k\frac{2\pi}{T_s} \right) d\lambda \tag{1.56}$$

in which the integration expression can be simplified using the impulse function property

$$\int_{-\infty}^{\infty} X_c(\lambda)\delta(\lambda_0 - \lambda)d\lambda = X_c(\lambda_0) \tag{1.57}$$

Fig. 1.54 $x_c(t)$ graph for
Example 1.15

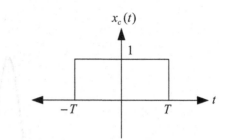

as follows

$$\frac{1}{2\pi}\int_{-\infty}^{\infty} X_c(\lambda)\delta\left(\frac{w}{T_s} - \lambda - k\frac{2\pi}{T_s}\right)d\lambda = \frac{1}{2\pi}X_c\left(\frac{w}{T_s} - k\frac{2\pi}{T_s}\right) \qquad (1.58)$$

Finally, when (1.58) is used in (1.56), we get the desired final expression as

$$X_n(w) = \frac{1}{T_s}\sum_{k=-\infty}^{\infty} X_c\left(\frac{w}{T_s} - k\frac{2\pi}{T_s}\right). \qquad (1.59)$$

Exercise: The inverse Fourier transform for digital signals is given as

$$x[n] = \frac{1}{2\pi}\int_{2\pi} X_n(w)e^{jwn}dw. \qquad (1.60)$$

Starting from the right hand side of (1.60) and replacing $X_n(w)$ in (1.60) by (1.59) obtain the left hand side of (1.60).

Example 1.15 The time domain signal given in Fig. 1.54 is to be sampled. Determine the sampling frequency such that the digital signal contains sufficient information about analog signal and analog signal can be reconstructed from the digital samples.

Solution 1.15 To determine the sampling frequency, we need to know the largest and smallest positive frequencies available in the signal spectrum. For this purpose, we calculate the Fourier transform of the continuous time signal and determine the largest and smallest positive frequencies available in the signal spectrum. The Fourier of the continuous time signal is computed as

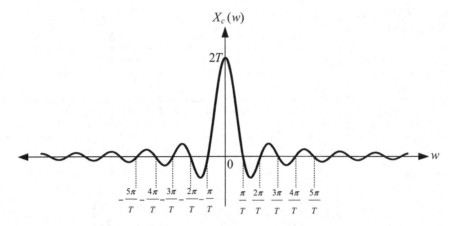

Fig. 1.55 Fourier transform of $x_c(t)$ in Fig. 1.54

$$X_c(w) = \int_{-\infty}^{\infty} x_c(t)e^{-jwt}dt$$

$$= \int_{-T}^{T} 1e^{-jwt}dt$$

$$= \frac{e^{jwT} - e^{-jwT}}{jw}$$

$$= \frac{2\sin(wT)}{w}$$

The graph of the Fourier transform is depicted in Fig. 1.55. Since

$$X_c(0) = \frac{0}{0}$$

the value of the Fourier transform at origin can be computed using the L'Hôpital's rule. If we take the derivatives of numerator and denominator of $X_c(w)$ w.r.t w and evaluate it for $w = 0$, we obtain

$$\left.\frac{dX_c(w)}{dw}\right|_{w=0} = \left.\frac{2T\cos(wT)}{1}\right|_{w=0} \rightarrow \left.\frac{dX_c(w)}{dw}\right|_{w=0} = 2T$$

which is nothing but the value of $X_c(w)$ at origin, i.e., $X_c(0)$.

As it is seen from Fig. 1.55, the largest positive frequency in the signal spectrum goes to infinity and the smallest non-negative frequency is 0. We need to choose infinity as sampling frequency and this is not a feasible value for practical implementations. However, as it is seen from the Fourier transform graph, the amplitude of the signal spectrum decreases sharply when frequency is beyond $\frac{\pi}{T}$. So, we can assume that the spectrum amplitude is negligible beyond a frequency value. We can choose the largest frequency as $w_N = \frac{4\pi}{T}$, and according to the chosen frequency, we can write the lower bound for sampling frequency as

$$w_s > 2w_N \rightarrow w_s > 2\frac{4\pi}{T}$$

$$w_s > \frac{8\pi}{T} \rightarrow \frac{2\pi}{T_s} > \frac{8\pi}{T} \rightarrow T_s < \frac{T}{4}$$

$$f_s > \frac{4}{T}$$

Let's assume that the sampling period is chosen as $T_s = \frac{T}{8}$. This means that we take $\frac{2T}{\frac{T}{8}} = 16$ samples from rectangle signal per second. And these 16 samples are sufficient for reconstruction of the rectangle signal.

1.4.2 Drawing the Frequency Response of Digital Signal in Case of Aliasing (Practical Method)

In sampling operation if the sampling frequency is chosen as

$$f_s < 2w_N$$

where w_N is the bandwidth of the low pass analog signal, then aliasing occurs in Fourier transform of the digital signal $x[n]$, i.e., in graph of $X_n(w)$. The relations between digital signal and continuous time signal in time and frequency domains are as

$$x[n] = x_c(t)|_{t=nT_s} \rightarrow x[n] = x_c(nT_s)$$

$$X_n(w) = \frac{1}{T_s} \sum_{k=-\infty}^{\infty} X_c\left(\frac{w}{T_s} - kw_s\right).$$

Let the Fourier transform of the continuous time signal to be sampled be as in Fig. 1.56.

If $f_s < 2w_N$, then the graph of $\frac{1}{T_s}X_c\left(\frac{w}{T_s}\right)$ happens to be as in Fig. 1.57.

If Fig. 1.57 is inspected carefully it is seen that when $f_s < 2w_N$, the function $\frac{1}{T_s}X_c\left(\frac{w}{T_s}\right)$ takes values outside the interval $(-\pi, \pi)$ on horizontal axis. In Fig. 1.58, the shadowed triangles denoted by 'A' and 'B' show the intervals outside $(-\pi, \pi)$ where the function $\frac{1}{T_s}X_c\left(\frac{w}{T_s}\right)$ has nonzero value.

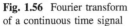

Fig. 1.56 Fourier transform of a continuous time signal

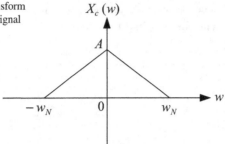

Fig. 1.57 Graph of $\frac{1}{T_s} X_c\left(\frac{w}{T_s}\right)$

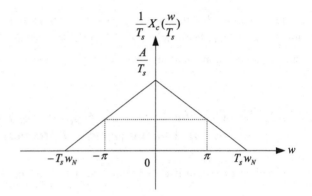

Fig. 1.58 The graph of $\frac{1}{T_s} X_c\left(\frac{w}{T_s}\right)$

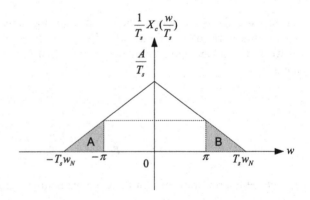

If the shadowed triangles 'A' and 'B' in Fig. 1.58 are shifted to the right and left by 2π, we obtain the graphic in Fig. 1.59.

If the overlapping lines in Fig. 1.59 are summed, we obtain the graphic shown in bold lines in Fig. 1.60. As it is clear from Fig. 1.60, due to the overlapping regions the original signal is spectrum is destroyed.

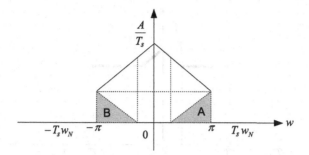

Fig. 1.59 Shifting of the shadowed triangles

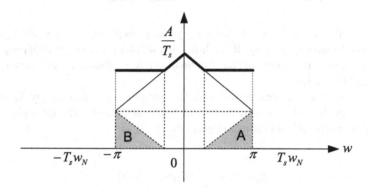

Fig. 1.60 Summation of the overlapping lines

The amount of this destruction depends on the widths of the shadowed triangles. In other words, as the function $\frac{1}{T_s}X_c\left(\frac{w}{T_s}\right)$ extends outside the interval $(-\pi, \pi)$ more, the amount of distortion on the original signal due to overlapping increases.

The graph obtained after summing the overlapping lines is depicted alone in Fig. 1.61.

Let's now, step by step, describe drawing the graph of $X_n(w)$ in case of aliasing in an easy and practical manner.

Step 1: First we draw the graph of $\frac{1}{T_s}X_c\left(\frac{w}{T_s}\right)$. For this purpose, we divide the horizontal axis of the graph of $X_c(w)$ by $1/T_s$ i.e., we multiply the horizontal axis by T_s, and multiply the amplitude values by $1/T_s$.

Step 2: If the sampling frequency is chosen as $f_s < 2w_N$, then aliasing occurs in the Fourier transform of $x[n]$, i.e., aliasing occurs in $X_n(w)$. And in this case, the graph of $\frac{1}{T_s}X_c\left(\frac{w}{T_s}\right)$ extends beyond the interval $(-\pi, \pi)$. The portion of the graph extending to the left of $-\pi$ is denoted by 'A', and the potion extending to the right of π is denoted by 'B'.

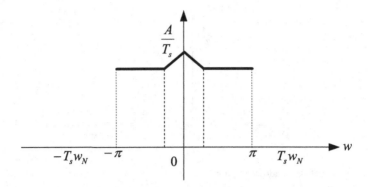

Fig. 1.61 The resulting graph after summing the overlapping lines

Step 3: The portion of the graph denoted by 'A' in Step 2 is shifted to the right by 2π, and the portion denoted by 'B' is shifted to the left by 2π. The overlapping lines are summed and one period of $X_n(w)$ around origin is obtained. Let's denote this one period by $X_{n1}(w)$.

Step 4: In the last step, one period of $X_n(w)$ around origin denoted by $X_{n1}(w)$ is shifted to the left and right by multiples of 2π and all the shifted replicas are summed to get $X_n(w)$, this is mathematically stated as

$$X_n(w) = \sum_{k=-\infty}^{\infty} X_{n1}(w - k2\pi).$$

Example 1.16 The Fourier transform of continuous time signal $x_c(t)$ is shown in Fig. 1.62. This signal is sampled and digital signal $x[n] = x_c(t)|_{t=nT_s} \rightarrow x[n] = x_c(nT_s)$, $T_s = 1/64$ is obtained. Draw the graph of the Fourier transform digital signal, i.e., draw the graph of $X_n(w)$.

Solution 1.16

Step 1: First we draw the graph of $\frac{1}{T_s}X_c\left(\frac{w}{T_s}\right)$, for this purpose, we multiply the horizontal axis of $X_c(w)$ in Fig. 1.62 by $T_s = 1/64$ and multiply the vertical axis of $X_c(w)$ in Fig. 1.62 by $1/T_s = 64$. The resulting graph is shown in Fig. 1.63.

Fig. 1.62 Fourier transform of a low pass input signal

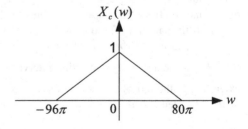

Fig. 1.63 The graph of $\frac{1}{T_s} X_c\left(\frac{w}{T_s}\right)$

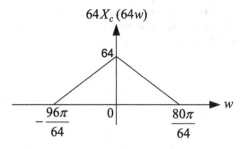

Fig. 1.64 The graph of $\frac{1}{T_s} X_c\left(\frac{w}{T_s}\right)$

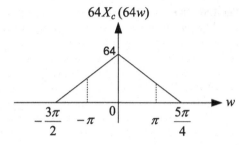

Fig. 1.65 The portions of graph outside $(-\pi, \pi)$ interval are labelled by 'A' and 'B'

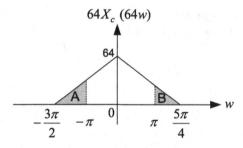

The graph in Fig. 1.63 is drawn more in details as in Fig. 1.64 where we see that the graph extends to the outside of the $(-\pi, \pi)$ interval. And in fact, the parts of the $\frac{1}{T_s} X_c\left(\frac{w}{T_s}\right)$ extending beyond $(-\pi, \pi)$ cause the spectral overlapping problem due to the 2π periodicity of $X_n(w)$.

Step 2: We shadow the portion of the graphs outside the $(-\pi, \pi)$ interval and denote them by the letters 'A' and 'B', we obtain the graph in Fig. 1.65.

If the shadowed portions labelled by 'A' and 'B' are shifted to the right and to the left by 2π, we obtain the graph in Fig. 1.66.

In Fig. 1.66, we can write the equations of the overlapping lines for the interval $(-\pi, -3\pi/4)$ as $\frac{128}{3\pi} w + 64$ and $-\frac{256}{5\pi} w - \frac{192}{5}$, and when these two equations are summed, we obtain $-\frac{128}{15\pi} w + \frac{128}{5}$. In a similar manner, if we write the equations of the overlapping lines for the interval $(\pi/2, \pi)$ and sum them, we obtain

Fig. 1.66 Shadowed portions
are shifted to the right and to
the left by 2π

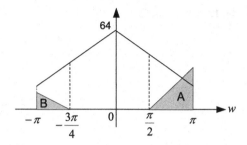

Fig. 1.67 One period of
$X_n(w)$ around origin

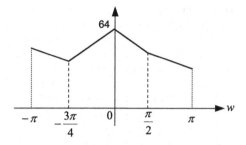

Fig. 1.68 Fourier transform
of a continuous time signal

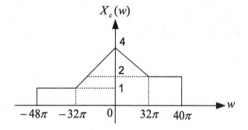

$-\frac{128}{15\pi}w + \frac{128}{3}$. After summing the overlapping line equations, we can draw one
period of $X_n(w)$ around origin as in Fig. 1.67.

Step 3: In the last step, we shift one period of $X_n(w)$ around origin to the left and
right by multiples of 2π and summing all the non-overlapping shifted replicas, we
obtain the graph of $X_n(w)$.

Exercise: The Fourier transform of a continuous time signal $x_c(t)$ is depicted in
Fig. 1.68.

This signal is sampled with sampling period $T_s = 1/32$ and digital signal $x[n]$ is
obtained. Draw the Fourier transform of $x[n]$.

Exercise: The Fourier transform of a continuous time signal $x_c(t)$ is depicted in
Fig. 1.69.

This signal is sampled with sampling period $T_s = 1/32$ and digital signal $x[n]$ is
obtained. Draw the Fourier transform of $x[n]$.

Fig. 1.69 Fourier transform
of a continuous time signal

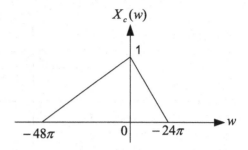

1.5   Reconstruction of an Analog Signal from Its Samples

To obtain a digital signal $x[n]$ from an analog signal $x_c(t)$ via sampling operation, we first multiply the analog signal by an impulse train $s(t)$ and obtain the product signal $x_s(t) = x_c(t)s(t)$. Then we collect the amplitude values of impulses from $x_s(t)$ and form the digital sequence $x[n]$.

Now we wonder the reverse operation, i.e., assume that we have the digital sequence $x[n]$, then how can we construct the analog signal $x_c(t)$? To achieve this, we will just follow the reverse operations. That is, we will first obtain $x_s(t)$ from $x[n]$, then from $x_s(t)$ we will extract $x_c(t)$.

Let's study the reconstruction operation in time domain as shown in Fig. 1.70.

As it is depicted in Fig. 1.70, we can write mathematical expression for the product signal $x_s(t)$ in terms of the elements of digital signal $x[n]$ but we have no way to write an expression for $x_c(t)$ using $x_s(t)$. Hence, we cannot solve the reconstruction problem in time domain. Let's inspect the reconstruction operation in frequency domain then. Assume that $x_c(t)$ is a low pass signal and its Fourier transform is as given in Fig. 1.71.

Considering the Fourier transform in Fig. 1.71, we can draw the Fourier transform of the product signal $x_s(t)$ as in Fig. 1.72. The Fourier transform of $x_s(t)$ is a periodic signal with period w_s and it's one period around origin equals to $\frac{1}{T_s}X_c(w)$ in case of no aliasing.

It is clear from Fig. 1.72 that for no aliasing, we should have

$$w_s > 2w_N \rightarrow \frac{2\pi}{T_s} > 2w_N \tag{1.61}$$

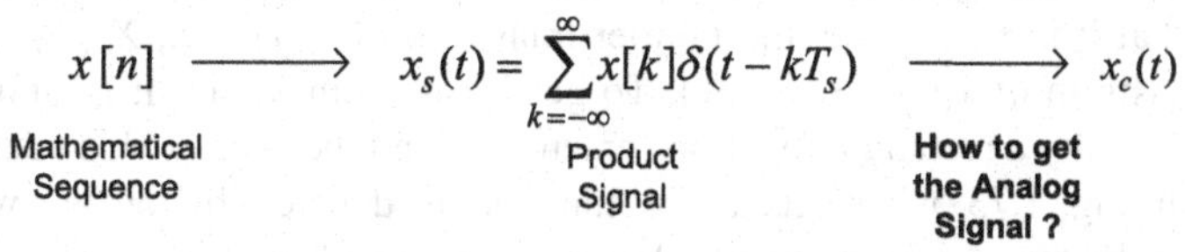

$$x[n] \quad \longrightarrow \quad x_s(t) = \sum_{k=-\infty}^{\infty} x[k]\delta(t - kT_s) \quad \longrightarrow \quad x_c(t)$$

Mathematical   Product   How to get
Sequence   Signal   the Analog
                                                Signal ?

Fig. 1.70 Reconstruction operation in time domain

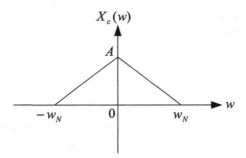

Fig. 1.71 Fourier transform of $x_c(t)$

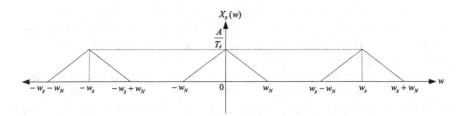

Fig. 1.72 Fourier transform of $x_s(t)$

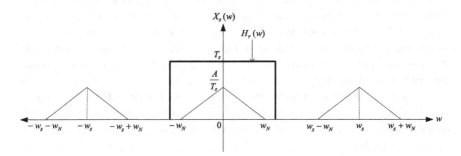

Fig. 1.73 Multiplication of $X_s(w)$ by rectangle function $H_r(w)$

$$\frac{2\pi}{T_s} > 2w_N \rightarrow w_N < \frac{\pi}{T_s}. \tag{1.62}$$

Now consider the reconstruction operation in frequency domain. We had problem in converting $x_s(t)$ to $x_c(t)$ in time domain. However, it is clear from Fig. 1.72 that it is easy to get the Fourier transform of $x_c(t)$, i.e., $X_c(w)$ from the Fourier transform of $x_s(t)$, i.e., $X_s(w)$. To get $X_c(w)$ from $X_s(w)$, it is sufficient to multiply $X_s(w)$ by a rectangle function centered around the origin. This operation is depicted in Fig. 1.73 where rectangle function is denoted by $H_r(w)$ which is nothing but the transfer function of a low pass analog filter.

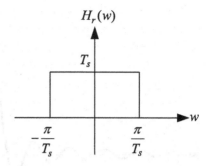

The Fourier transform of the low pass analog filter is depicted in Fig. 1.74 alone. In fact, the filter under consideration is an ideal lowpass filter, and it is used just to illustrate the reconstruction operation. In practice, such ideal filters are not available, and practical non-ideal filters are employed for reconstruction operations.

The time domain expression of the analog filter with the frequency response depicted in Fig. 1.74 can be calculated using the inverse Fourier transform formula as follows:

$$h_r(t) = \frac{1}{2\pi} \int_{-\infty}^{\infty} H_r(w)e^{jwt} dw$$

$$= \frac{1}{2\pi} \int_{-\frac{\pi}{T_s}}^{\frac{\pi}{T_s}} T_s e^{jwt} dw$$

$$= \frac{T_s}{2\pi} e^{jwt} \Big|_{-\frac{\pi}{T_s}}^{\frac{\pi}{T_s}}$$

$$= \frac{T_s}{j2\pi t} \left(e^{j\frac{\pi}{T_s}t} - e^{-j\frac{\pi}{T_s}t} \right)$$

where using the property $\sin(\theta) = \frac{1}{2j}\left(e^{j\theta} - e^{-j\theta}\right)$, we obtain

$$h_r(t) = \frac{\sin\left(\frac{\pi t}{T_s}\right)}{\frac{\pi t}{T_s}}. \tag{1.63}$$

Since

$$\sin c(x) = \frac{\sin(\pi x)}{\pi x} \tag{1.64}$$

the mathematical expression in (1.63) can be written in terms of $\sin c(\cdot)$ function as

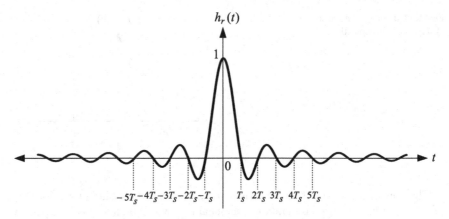

Fig. 1.75 Reconstruction filter impulse response

$$h_r(t) = \sin c\left(\frac{t}{T_s}\right). \tag{1.65}$$

The graph of the reconstruction filter $h_r(t)$ is depicted in Fig. 1.75 where it is clear that the reconstruction filter takes 0 value at every multiple of T_s.

As we explained before the Fourier transform of the continuous time signal can be written as the multiplication of $X_s(w)$ and $H_r(w)$ i.e.,

$$X_c(w) = X_s(w)H_r(w). \tag{1.66}$$

Since multiplication in frequency domain equals to convolution in time domain, (1.66) can be also be expressed as

$$x_c(t) = x_s(t) * h_r(t) \tag{1.67}$$

where substituting

$$\sum_{n=-\infty}^{\infty} x[n]\delta(t - nT_s)$$

for $x_s(t)$, we obtain

$$\begin{aligned} x_c(t) &= \sum_{n=-\infty}^{\infty} x[n]\delta(t - nT_s) * h_r(t) \\ &= \sum_{n=-\infty}^{\infty} x[n]h_r(t - nT_s) \end{aligned} \tag{1.68}$$

which is nothing but the reconstruction expression of the analog signal $x_c(t)$.

Note: $f(t) * \delta(t - t_0) = f(t - t_0)$

Using (1.63) in (1.68) the reconstructed analog signal from its samples can be written as

$$x_c(t) = \sum_{n=-\infty}^{\infty} x[n] \frac{\sin\left(\pi \frac{(t-nT_s)}{T_s}\right)}{\frac{\pi(t-nT_s)}{T_s}} \tag{1.69}$$

or in terms of $\sin c(\cdot)$ function, it is written as

$$x_c(t) = \sum_{n=-\infty}^{\infty} x[n] \sin c\left(\frac{t - nT_s}{T_s}\right) \tag{1.70}$$

Example 1.17 The continuous time signal $x_c(t) = \sin(2\pi t)$ is sampled by sampling period $T_s = \frac{1}{4}$ s.

(a) Write the digital sequence $x[n]$ obtained after sampling operation.
(b) Assume that $x[n]$ is transmitted and available at the receiver. Reconstruct the analog signal at the receiver side from its samples, i.e., using $x[n]$ reconstruct the analog signal $x_c(t)$.

Solution 1.17

(a) The frequency of the sinusoidal signal $x_c(t) = \sin(2\pi t)$ is 1 Hz, and its period is 1 s. Sampling period is $T_s = \frac{1}{4}$ s. Every multiple of T_s, we take a sample from the sinusoidal signal. The graph of the sinusoidal signal and the samples taken from its one period are indicated in Fig. 1.76.

Since sampling frequency is $f_s = 4$ Hz, we take 4 samples per-second from the signal. The samples taken from one period of the sinusoidal signal can be written as $[0 \quad 1 \quad 0 \quad -1]$. Since the sine signal is defined from $-\infty$ to ∞. The obtained digital signal is a periodic signal and in this digital signal, the repeating pattern happens to be $[0 \quad 1 \quad 0 \quad -1]$. The digital signal obtained from the sampling operation can be written as

Fig. 1.76 Sampling of sine signal

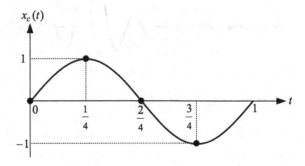

$$x[n] = \begin{bmatrix} \cdots & 0 & 1 & 0 & -1 & \overset{n=0}{\overset{\frown}{0}} & 1 & 0 & -1 & 0 & 1 & 0 & -1 & \cdots \end{bmatrix}$$

$\underbrace{\hphantom{0 \quad 1 \quad 0 \quad -1}}_{\text{Repeating pattern}}$

$$(1.71)$$

(b) At the receiver, the analog signal can be reconstructed from its samples using

$$x_c(t) = \sum_{n=-\infty}^{\infty} x[n] h_r(t - nT_s) \qquad (1.72)$$

where $T_s = \frac{1}{4}$ and

$$h_r(t) = \frac{\sin\left(\frac{\pi t}{T_s}\right)}{\frac{\pi t}{T_s}}. \qquad (1.73)$$

Using the $x[n]$ in (1.72), the reconstructed signal can be written as

$$x_c(t) = \cdots + h_r(t + 3T_s) - h_r(t + T_s) + h_r(t - T_s) - h_r(t - 3T_s) + \cdots \qquad (1.74)$$

The graph of $h_r(t)$ in (1.73) is depicted in Fig. 1.77 where it is clear that the amplitude of the main lobe of $h_r(t)$ equals to 1, and the function equals to 0 when t is a multiple of T_s.

The shifted copies of $h_r(t)$ and their summation is illustrated in Fig. 1.78.

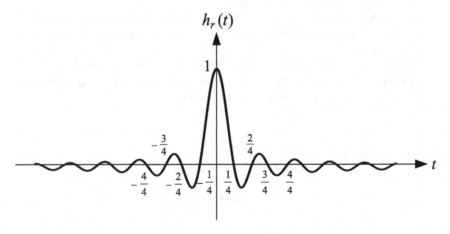

Fig. 1.77 Reconstruction filter impulse response

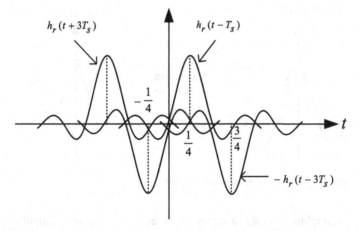

Fig. 1.78 Summing the shifted $\sin c(\cdot)$ functions to reconstruct the analog signal

If we only pay attention to the main lobes in Fig. 1.78, we see that the reconstruction signal resembles to the sine signal. Overlapping tails improve the accuracy of the reconstructed signal.

1.5.1 Approximation of the Reconstruction Filter

The reconstruction filter $h_r(t) = \sin c\left(\frac{t}{T_s}\right)$ is depicted in Fig. 1.79 where it is seen that the filter has a large main lobe and small side lobes, and as the time values

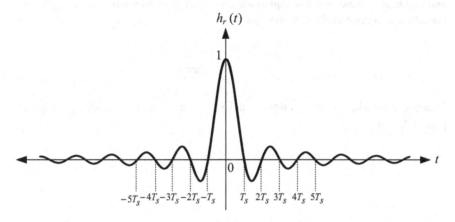

Fig. 1.79 Reconstruction filter impulse response

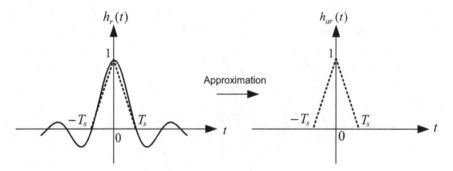

Fig. 1.80 Approximation of the reconstruction filter

increase, the amplitudes of the side lobes decrease. To construct a simplified model for the reconstruction filter, we can approximate the lobes by isosceles triangles.

In Fig. 1.80 the main lobe of the reconstruction filter is approximated by an isosceles triangle and the side lobes are all omitted. This type of approximation can also be called as linear approximation.

For the triangle in Fig. 1.80, we can write line equations for the left and right edges. For the left edge, the line equation is

$$\frac{t}{T_s} + 1, \quad -T_s \le t < 0,$$

for the right edge, the line equation is

$$-\frac{t}{T_s} + 1, \quad 0 \le t \le T_s$$

and combining these two line equations into a single expression, we can write the linearly approximated filter expression as

$$h_{ar} = \begin{cases} -\frac{|t|}{T_s} + 1 & 0 \le |t| \le T_s \\ 0 & otherwise \end{cases}.$$

Example 1.18 The continuous time signal $x_c(t) = \sin(2\pi t)$ is sampled by sampling period $T_s = \frac{1}{4}$.

(a) Write the digital sequence $x[n]$ obtained after sampling operation.
(b) Assume that $x[n]$ is transmitted and available at the receiver. Reconstruct the analog signal at the receiver side from its samples using approximated reconstruction filter.

Solution 1.18

(a) We solved this problem before and found the digital signal as

$$x[n] = \left[\cdots \ 0 \ 1 \ 0 \ -1 \ \overset{\overset{n=0}{\frown}}{0} \ \underbrace{1 \ 0 \ -1}_{\text{Repeating pattern}} \ 0 \ 1 \ 0 \ -1 \ \cdots \right].$$

$$(1.75)$$

(b) At the receiver side, the analog signal can be reconstructed from its samples using

$$x_c(t) = \sum_{n=-\infty}^{\infty} x[n] h_{ar}(t - nT_s) \qquad (1.76)$$

where $T_s = \frac{1}{4}$ and $h_{ar}(t)$ is the approximated reconstruction filter. Using the $x[n]$ found in the previous part, and expanding (1.76), the reconstructed signal can be written as

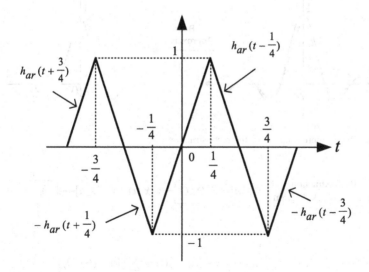

Fig. 1.81 Reconstruction of analog signal using approximated filter

$$x_c(t) = \cdots + h_{ar}(t+3T_s) - h_{ar}(t+T_s) + h_{ar}(t-T_s) - h_{ar}(t-3T_s) + \cdots$$

$$(1.77)$$

The shifted copies of $h_{ar}(t)$ in (1.77) and their summation is illustrated in Fig. 1.81.

As it is seen from Fig. 1.81, the reconstructed signal resembles to the sine signal.

Now we ask the question: How can we obtain a better reconstructed sine signal?

Answer

Either we can use a better approximated filter or take more samples from one period of the signal, i.e., increase the sampling frequency which means, decrease the sampling period. To get a better approximated filter, we can represent the side-lobes by the small triangles.

A better approximation of the reconstruction filter is illustrated in Fig. 1.82 where it is seen that two side lobes are approximated by triangles. Although improved linear approximation improves the accuracy of the reconstructed signal, the sharp discontinuities of the linear approximated filter makes the realization of the filter difficult.

Reconstruction operation can be illustrated using block diagrams as in Fig. 1.83.

Fig. 1.82 Better approximation of the reconstruction filter

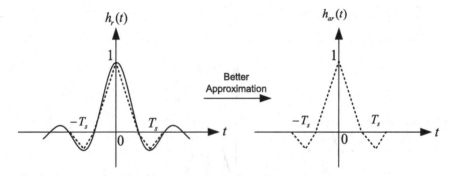

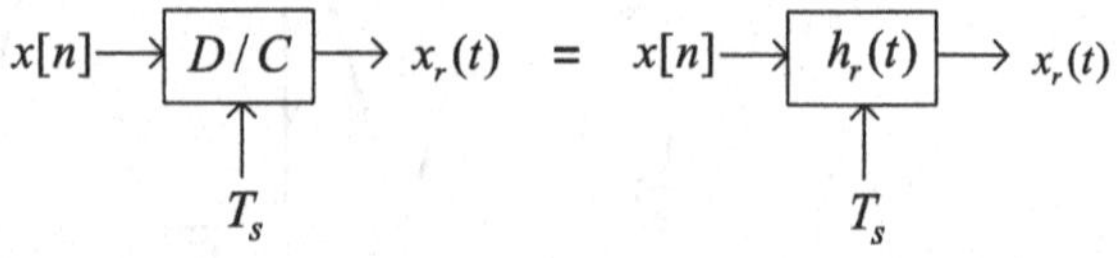

Fig. 1.83 Reconstruction operation using block diagram

$$x_r(t) = \sum_{n=-\infty}^{\infty} x[n]h_r(t-nT_s)$$

In Fig. 1.83, if $h_r(t) = \sin c\left(\frac{t}{T_s}\right)$, then perfect reconstruction occurs, i.e., $x_r(t) = x_c(t)$.

1.6 Discrete Time Processing of Continuous Time Signals

Currently most of the electronic devices are produced using digital technology. For this reason, analog signals are usually converted to digital signals and processed by digital electronic systems. These electronic units can be digital filters, equalizers, amplifiers, etc. In Fig. 1.84, the general system for digital processing of analog system is depicted.

The system in Fig. 1.84 can be inspected both in time and frequency domains assuming that discrete time system is linear and time invariant. Let's first write the relations among signals in time, and then in frequency domain.

Time Domain Relations:

$$x[n] = x_c(nT_{s1})y[n] = x[n] * h[n] \quad y_r(t) = \sum_{n=-\infty}^{\infty} y[n]h_r(t - nT_{s2}) \qquad (1.78)$$

If perfect reconstruction filter is to be employed, then

$$h_r(t) = \sin c\left(\frac{t}{T_{s2}}\right). \qquad (1.79)$$

Frequency Domain Relations:

$$X_n(w) = \frac{1}{T_{s1}} \sum_{k=-\infty}^{\infty} X_c\left(\frac{w}{T_{s1}} - kw_{s1}\right) \qquad (1.80)$$

where

$$w_{s1} = \frac{2\pi}{T_{s1}}, \quad Y_n(w) = X_n(w)H_n(w). \qquad (1.81)$$

To write the frequency domain relation between $y[n]$ and $y_r(t)$, let's remember the two-stage reconstruction process illustrated as follows

Fig. 1.84 Digital processing of a continuous time signal

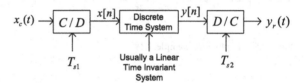

$$y[n] \longrightarrow \boxed{\begin{array}{c} \text{Convert mathematical} \\ \text{sequence to impulse} \\ \text{train} \end{array}} \longrightarrow y_s(t) = \sum_{n=-\infty}^{\infty} y[n]\delta(t-nT_{s2}) \longrightarrow \boxed{h_r(t)} \longrightarrow y_r(t)$$

$$T_{s2}$$

$$y[n] = y_c(nT_{s2})$$

We have

$$Y_s(w) = \frac{1}{T_{s2}} \sum_{k=-\infty}^{\infty} Y_c(w-kw_{s2}) \quad Y_n(w) = \frac{1}{T_{s2}} \sum_{k=-\infty}^{\infty} Y_c\left(\frac{w}{T_{s2}}-kw_{s2}\right) \quad (1.82)$$

$$Y_n(w) = Y_s\left(\frac{w}{T_{s2}}\right) \rightarrow Y_r(w) = H_r(w)Y_s(w) \rightarrow Y_r(w) = H_r(w)Y_n(T_{s2}w). \quad (1.83)$$

By combining $X_n(w) = \frac{1}{T_s} \sum_{k=-\infty}^{\infty} X_c\left(\frac{w}{T_s}-kw_s\right), Y_n(w) = X_n(w)H_n(w)$ and $Y_r(w) = H_r(w)Y_n(T_{s2}w)$, we get the relation between $Y_r(w)$ and $X_c(w)$ as

$$Y_r(w) = H_r(w)H_n(T_{s2}w)\frac{1}{T_{s1}} \sum_{k=-\infty}^{\infty} X_c\left(\frac{T_{s2}}{T_{s1}}w-kw_{s1}\right) \quad (1.84)$$

If $T_{s1} = T_{s2} = T_s$, then (1.84) reduces to

$$Y_r(w) = \begin{cases} T_s H_n(T_s w)X_c(w), & -\frac{\pi}{T_s} \leq w \leq \frac{\pi}{T_s} \\ 0, & otherwise \end{cases}. \quad (1.85)$$

Note: $H_r(w) = \begin{cases} T_s & if -\frac{\pi}{T_s} \leq w \leq \frac{\pi}{T_s} \\ 0 & otherwise \end{cases}$

Example 1.19 In Fig. 1.85, the graphs of $X_c(w)$ and $X_n(w)$ are depicted. In addition, $x[n] = x_c(t)|_{t=nT_s}$. By comparing the graphs of $X_c(w)$ and $X_n(w)$, write $X_c(w)$ in terms of $X_n(w)$.

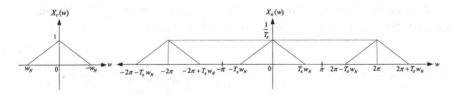

Fig. 1.85 Graphs for Example 1.19

Fig. 1.86 One period of $X_n(w)$ around origin

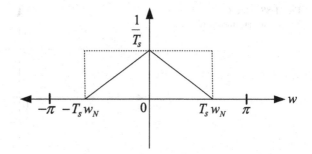

Solution 1.19 First let's write the expression for one period of $X_n(w)$ around origin as

$$X_n(w) \quad -\pi \leq w < \pi \tag{1.86}$$

which is graphically shown as in Fig. 1.86.

If we divide the horizontal axis of $X_n(w)$ by T_s, we get

$$X_n(T_sw) \quad -\frac{\pi}{T_s} \leq w < \frac{\pi}{T_s} \tag{1.87}$$

which is graphically depicted in Fig. 1.87.

If we multiply the amplitudes by T_s, we obtain

$$T_sX_n(T_sw) \quad -\frac{\pi}{T_s} \leq w < \frac{\pi}{T_s}. \tag{1.88}$$

which is graphically depicted in Fig. 1.88.

Figure 1.88 is nothing but the graph of $X_c(w)$. As a result, we can conclude that if $x[n] = x_c(t)|_{t=nT_s}$, then we can express Fourier transform of $x_c(t)$ i.e., $X_c(w)$ in terms of Fourier transform of $x[n]$ i.e., $X_n(w)$ as

$$X_c(w) = T_sX_n(T_sw) \quad -\frac{\pi}{T_s} \leq w < \frac{\pi}{T_s}. \tag{1.89}$$

Fig. 1.87 One period of $X_n(T_sw)$ around origin

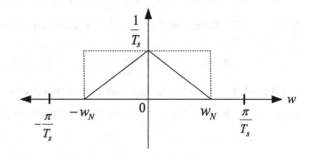

Fig. 1.88 One period of
$T_s X_n(T_s w)$ around origin

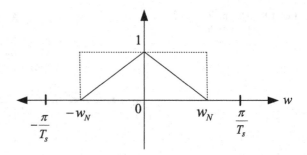

Fig. 1.89 Continuous to
digital converter

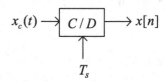

Example 1.20 For the continuous to digital converter given in Fig. 1.89, assume
that the sampling frequency is high enough so that there is no aliasing in frequency
domain. $X_n(w)$ is the Fourier transform of $x[n]$, and $X_c(w)$ is the Fourier transform
of $x_c(t)$. Write one period of $X_n(w)$ in terms of $X_c(w)$.

Solution 1.20 Since $X_n(w)$ is the Fourier transform of a digital signal, $X_n(w)$ is
periodic and its period equals 2π, the relation between $X_n(w)$ and $X_c(w)$ is given as

$$X_n(w) = \frac{1}{T_s} \sum_{k=-\infty}^{\infty} X_c\left(\frac{w}{T_s} - k\frac{2\pi}{T_s}\right) \qquad (1.90)$$

which is written explicitly as

$$X_n(w) = \cdots + \underbrace{\frac{1}{T_s}X_c\left(\frac{w}{T_s} + \frac{2\pi}{T_s}\right)}_{n=-1} + \underbrace{\frac{1}{T_s}X_c\left(\frac{w}{T_s}\right)}_{n=0} + \underbrace{\frac{1}{T_s}X_c\left(\frac{w}{T_s} - \frac{2\pi}{T_s}\right)}_{n=1} + \cdots \quad (1.91)$$

In (1.91), let $Y_c(w) = \frac{1}{T_s}X_c\left(\frac{w}{T_s}\right)$, then it is obvious that $Y_c(w - 2\pi) = \frac{1}{T_s}X_c\left(\frac{w}{T_s} - \frac{2\pi}{T_s}\right)$. The explicit expression of $X_n(w)$ can be written as

$$X_n(w) = \cdots + Y_c(w + 2\pi) + Y_c(w) + Y_c(w - 2\pi) + \cdots \; ' \qquad (1.92)$$

From (1.92), it is obvious that one period of $X_n(w)$ is $Y_c(w)$, that is to say, one
period of $X_n(w)$ is $\frac{1}{T_s}X_c\left(\frac{w}{T_s}\right)$ and this can mathematically be written as

$$X_n(w) = \frac{1}{T_s} X_c\left(\frac{w}{T_s}\right) \qquad -\pi \le w < \pi \tag{1.93}$$

which can also be written as

$$T_s X_n(wT_s) = X_c(w) \qquad -\frac{\pi}{T_s} \le w < \frac{\pi}{T_s}. \tag{1.94}$$

Example 1.21 For the digital to continuous converter given in Fig. 1.90, let $Y_n(w)$ be the Fourier transform of $y[n]$. Because $Y_n(w)$ is the Fourier transform of a digital signal, it is periodic and its period equals 2π. Let $Y_{nop}(w)$ be the one period of $Y_n(w)$ around origin. That is $Y_{nop}(w) = Y_n(w)$ $-\pi \le w < \pi$. Write the Fourier transform of $y_r(t)$, i.e., $Y_r(w)$ in terms of $Y_{nop}(w)$.

Solution 1.21 Digital to continuous conversion operation is reminded in Fig. 1.91. As a result, we can write the relation between one period of $Y_n(w)$ and $Y_r(w)$ as

$$Y_r(w) = T_s Y_{nop}(T_s w) \tag{1.95}$$

The expression in (1.95) can also be written as

$$Y_r(w) = T_s Y_n(T_s w) \qquad -\frac{\pi}{T_s} \le w \le \frac{\pi}{T_s}. \tag{1.96}$$

$$y[n] \rightarrow \boxed{D/C} \rightarrow y_r(t)$$
$$\uparrow$$
$$T_s$$

Fig. 1.90 Digital to continuous converter

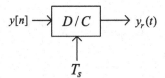

$$y[n] \longrightarrow y_s(t) \longrightarrow \boxed{h_r(t)} \longrightarrow y_r(t)$$

$$Y_n(w) = \sum_k Y_{nop}(w - k2\pi) \qquad Y_s(w) = Y_n(T_s w) \qquad Y_r(w) = T_s Y_{sop}(w)$$

$$Y_{nop} \qquad\qquad Y_{sop}(w) = Y_{nop}(T_s w) \qquad Y_r(w) = T_s Y_{nop}(T_s w)$$

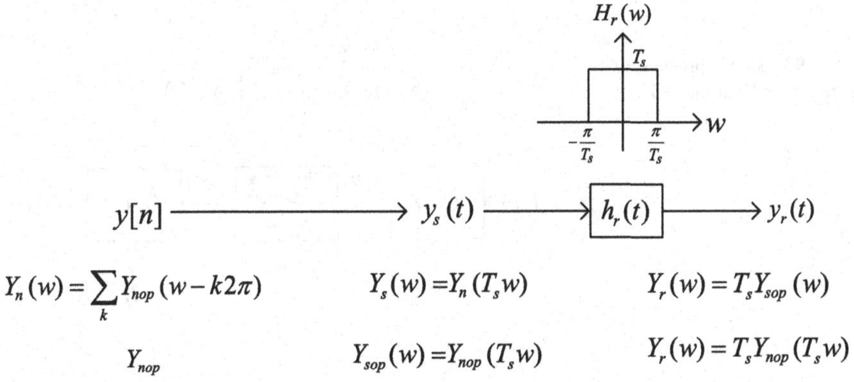

Fig. 1.91 Digital to continuous conversion

Example 1.22 If $x[n] = T_s y_c(t)\big|_{t=nT_s}$, write one period of the Fourier transform of $x[n]$ in terms of Fourier transform of $y_c(t)$. Assume that there is no aliasing.

Solution 1.22 Using the expression below

$$X_n(w) = \frac{T_s}{T_s} \sum_{k=-\infty}^{\infty} Y_c\left(\frac{w}{T_s} - k\frac{2\pi}{T_s}\right) \tag{1.97}$$

the relation in one period can be written as

$$X_n(w) = Y_c\left(\frac{w}{T_s}\right) \rightarrow Y_c(w) = X_n(T_s w). \tag{1.98}$$

Example 1.23 In Fig. 1.92 two signal processing systems are depicted. If both systems produce the same output $y_r(t)$ for the same input signal $x_c(t)$, find the relation between the impulse responses of continuous time and discrete time systems.

Solution 1.23 For the first system, the frequency domain relation between system input and output is

$$Y_r(w) = H_c(w)X_c(w) \tag{1.99}$$

Considering only one period (op) of the Fourier transforms of the digital signals around origin, the relations between input and output of each unit can be written as *C/D*:

$$X_{n-op}(w) = \frac{1}{T_s} X_c\left(\frac{w}{T_s}\right) \tag{1.100}$$

Disc.Time System:

$$Y_{n-op}(w) = H_n(w)X_{n-op}(w) \tag{1.101}$$

Fig. 1.92 Signal processing systems for Example 1.23

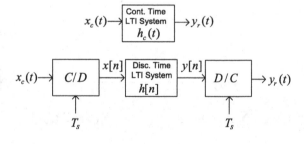

D/C:

$$Y_r(w) = T_s Y_{n-op}(T_s w) \tag{1.102}$$

If we combine the expressions (1.100–1.102), we get

$$Y_r(w) = H_{n-op}(T_s w) X_c(w). \tag{1.103}$$

If we equate the right hand sides of the Eqs. (1.99) and (1.103), we get

$$H_c(w) = H_{n-op}(T_s w) \rightarrow H_{n-op}(w) = H_c\left(\frac{w}{T_s}\right) \tag{1.104}$$

from which we can write the time domain relation for $h[n]$ and $h_c(t)$ as

$$h[n] = T_s h_c(t)|_{t=nT_s}. \tag{1.105}$$

1.7 Continuous Time Processing of Digital Signals

Digital signals can be processed by continuous time systems. For this purpose, the digital signal is first converted to continuous time signal then processed by a continuous time system whose output is back converted to a digital signal. The overall procedure is depicted in Fig. 1.93.

For the system in Fig. 1.93, time and frequency domain relations between block inputs and outputs are as follows:

Time domain relations are

$$x_c(t) = \sum_{n=-\infty}^{\infty} x[n] \sin c\left(\frac{t - nT_s}{T_s}\right) \quad y_c(t) = x_c(t) * h_c(t) \tag{1.106}$$

$$y[n] = y_c(t)|_{t=nT_s}. \tag{1.107}$$

Frequency domains relations are

$$X_c(w) = \begin{cases} T_s X_n(T_s w) & \text{if } -\frac{\pi}{T_s} \leq w \leq \frac{\pi}{T_s} \\ 0 & \text{otherwise} \end{cases} \quad Y_c(w) = X_c(w) H_c(w) \tag{1.108}$$

Fig. 1.93 Continuous time processing of digital signals

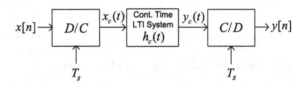

Fig. 1.94 Signal processing
units for Example 1.24

Fig. 1.95 Signal processing
units for Example 1.24

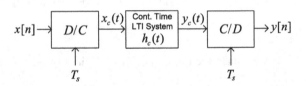

$$Y_n(w) = \frac{1}{T_s} \sum_{k=-\infty}^{\infty} Y_c\left(\frac{w}{T_s} - k\frac{2\pi}{T_s}\right). \tag{1.109}$$

Example 1.24 The signal processing units given in Figs. 1.94 and 1.95 have the same outputs for the same given inputs. Find the relation between the impulse responses of discrete and continuous time systems.

Solution 1.24 For the first system, the relation between input and output is

$$Y_n(w) = H_n(w)X_n(w). \tag{1.110}$$

Here $Y_n(w)$ is periodic with period 2π and one period of it can be written as either

$$Y_n(w) = H_n(w)X_n(w) \quad -\pi \leq w < \pi \tag{1.111}$$

or

$$Y_{n-op}(w) = H_n(w)X_{n-op}(w) \tag{1.112}$$

For the second system, the relations between block inputs and outputs are given as

$$X_c(w) = \begin{cases} T_s X_n(T_s w) & if -\frac{\pi}{T_s} \leq w \leq \frac{\pi}{T_s} \\ 0 & otherwise \end{cases} \quad Y_c(w) = X_c(w)H_c(w) \tag{1.113}$$

$$Y_n(w) = \frac{1}{T_s} \sum_{k=-\infty}^{\infty} Y_c\left(\frac{w}{T_s} - k\frac{2\pi}{T_s}\right). \tag{1.114}$$

If we combine the expressions in (1.113), we get

$$Y_c(w) = \begin{cases} H_c(w)T_s X_n(T_s w) & if -\frac{\pi}{T_s} \leq w \leq \frac{\pi}{T_s} \\ 0 & otherwise \end{cases} \tag{1.115}$$

and substituting (1.115) into (1.114), we obtain

$$Y_n(w) = \sum_{k=-\infty}^{\infty} H_c\left(\frac{w}{T_s} - k\frac{2\pi}{T_s}\right) X_n(w - k2\pi). \qquad (1.116)$$

One period of $Y_n(w)$ is

$$Y_{n-op}(w) = H_c\left(\frac{w}{T_s}\right) X_{n-op}(w) \qquad (1.117)$$

If we equate the right hand sides of (1.112) and (1.117)

$$H_n(w)X_{n-op}(w) = H_c\left(\frac{w}{T_s}\right) X_{n-op}(w) \rightarrow H_n(w) = H_c\left(\frac{w}{T_s}\right) \qquad (1.118)$$

which is can be expressed in time domain as

$$h[n] = T_s h_c(t)|_{t=nT_s}. \qquad (1.119)$$

Example 1.25 Sample continuous time signal in Fig. 1.96, and reconstruct the continuous time signal from its samples. Use triangle approximated reconstruction filter during reconstruction process.

Solution 1.25 The Fourier transform graph of a rectangle signal of length $2T$ around origin is repeated in Fig. 1.97.

For our example; $T = 2$, let's choose the approximate bandwidth of the rectangle pulse as $w_N = 2\pi/T \rightarrow w_N = 2\pi/2 \rightarrow w_N = \pi$. We can choose the sampling frequency according to

$$w_s > 2w_N \leftarrow w_s > 2\pi \qquad (1.120)$$

as

$$2\pi f_s > 2\pi \rightarrow f_s > 1 \rightarrow f_s = 2 \qquad (1.121)$$

which means that the sampling period is $T_s = \frac{1}{2}$. The sampling operation of the rectangle pulse is depicted in Fig. 1.98.

The digital sequence obtained after sampling of the rectangular signal is

Fig. 1.96 Continuous time signal for Example 1.25

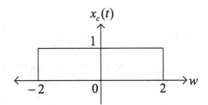

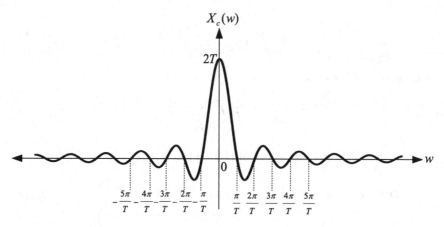

Fig. 1.97 Fourier transform of a rectangle signal

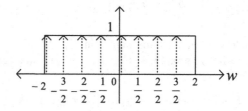

Fig. 1.98 Sampling of the rectangular signal

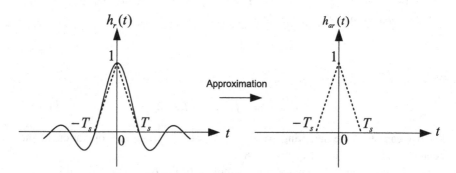

Fig. 1.99 Linear approximation of the reconstruction filter

$$x[n] = \begin{bmatrix} 1 & 1 & 1 & 1 & \underbrace{1}_{n=0} & 1 & 1 & 1 \end{bmatrix}. \tag{1.122}$$

The construction of the approximated reconstruction filter is repeated in Fig. 1.99.

Fig. 1.100 Linear approximation of the reconstruction filter for $T_s = \frac{1}{2}$

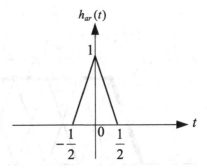

Note that our sampling period is $T_s = \frac{1}{2}$, then the approximated reconstruction filter becomes as in Fig. 1.100.

Now we can start the reconstruction operation, the reconstruction expression is given as

$$x_r(t) = \sum_{n=-\infty}^{\infty} x[n] h_{ar}(t - nT_s) \tag{1.123}$$

where $T_s = \frac{1}{2}$ s, and using our digital signal $x[n]$ and expanding the summation in (1.123), we obtain

$$
\begin{aligned}
x_r(t) = {} & h_{ar}\left(t + \frac{4}{2}\right) + h_{ar}\left(t + \frac{3}{2}\right) + h_{ar}\left(t + \frac{2}{2}\right) + h_{ar}\left(t + \frac{1}{2}\right) \\
& + h_{ar}(t) + h_{ar}\left(t - \frac{1}{2}\right) + h_{ar}\left(t - \frac{2}{2}\right) + h_{ar}\left(t - \frac{3}{2}\right).
\end{aligned} \tag{1.124}
$$

The shifted filters appearing in (1.124) is depicted in Fig. 1.101.

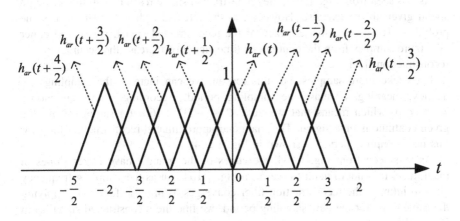

Fig. 1.101 Shifted triangle reconstruction filters

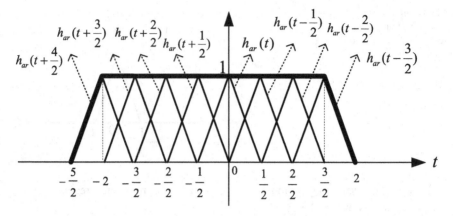

Fig. 1.102 Sum of the shifted reconstruction filters

Fig. 1.103 Reconstructed
signal for $T_s = \frac{1}{16}$

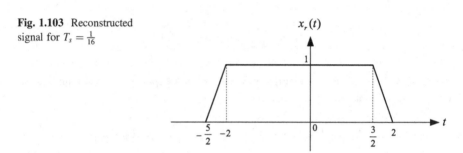

If the shifted graphs given in Fig. 1.101 are summed, we get the resulting graph shown in bold lines in Fig. 1.102.

In Fig. 1.103 the reconstructed signal is depicted alone.

As it is seen from Fig. 1.103, the reconstructed signal resembles to the rectangle signal given in the exercise. However, at the left and right sides we have some problems. To increase the accuracy of the reconstructed signal, we should either take more samples from the continuous time signal or increase the accuracy of the reconstruction filter.

Let's take more samples. For this reason, we can increase the sampling frequency, meaning, decrease the sampling period. Accordingly, we can choose $T_s = 1/16$, which means that we take $(2 - (-2)) \times 16 = 64$ samples from the given continuous time signal. The triangular approximated reconstruction filter for this new sampling period is shown in Fig. 1.104.

As it is seen from Fig. 1.104, the edges of the triangle have larger slopes in magnitude. It is not difficult to see from Fig. 1.104 that as the sampling frequency goes to infinity, the reconstruction filter converges to impulse function. Applying the same steps for the new sampling period, we find the reconstructed signal as in Fig. 1.105.

Fig. 1.104 Linear
approximation of the
reconstruction filter for $T_s = \frac{1}{2}$

Fig. 1.105 Reconstructed
signal for $T_s = \frac{1}{16}$

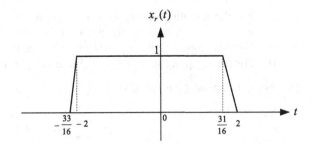

Fig. 1.106 Signal graph for
Example 1.26

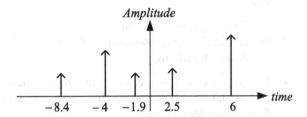

As it is seen from Fig. 1.105, we have a better reconstructed signal. Left and
right edges of the reconstructed signal have larger slopes.

Note: If unit is not provided for sampling period or for signal axis we accept it as
"second" by default.

Example 1.26 Is the signal given in Fig. 1.106 a digital signal?

Solution 1.26 Time axis of a digital signal consist of only integers. For the given
signal, real values appear along time axis. Hence, the signal is not a digital signal
but it is discrete amplitude continuous time signal. In fact the signal consists of
shifted impulses $\delta(t - t_0)$ which is a continuous function.

1.8 Problems

(1) For the sampling periods $T_s = 1$ s and $T_s = 1.5$ s, draw the graph of

$$s(t) = \sum_{n=-\infty}^{\infty} \delta(t - nT_s).$$

(2) The signal depicted in Fig. 1.107 is sampled.

   (a) For the sampling period $T_s = 1$ s, first draw the graph of impulse train function $s(t)$, then draw the graph of the product signal $x_s(t) = x_c(t)s(t)$. Find the digital signal $x[n]$ and draw its graph.
   (b) For the sampling period $T_s = 0.5$ s repeat part (a)

(3) For the impulse train function

$$s(t) = \sum_{n=-\infty}^{\infty} \delta(t - nT_s)$$

   find

   (a) Fourier series coefficients.
   (b) Fourier series representation.
   (c) Fourier transform.

(4) If $x_s(t) = x_c(t)s(t)$ where $s(t)$ is the impulse train and $x_c(t)$ is a continuous time signal, derive the Fourier transform expression of $x_s(t)$ in terms of the Fourier transform of $x_c(t)$.

(5) If $x[n] = x_c(nT_s)$, then derive the expression for the Fourier transform of $x[n]$ in terms of the Fourier transform of $x_c(t)$.

(6) Write mathematical equation for the lines depicted in Fig. 1.108, and then find the sum of these line equations.

(7) $x_c(t) = \cos(8\pi t)$ is sampled and $x[n] = x_c(nT_s)$ digital signal is obtained. According to this information, answer the following.

   (a) If the sampling period is $T_s = \frac{1}{4}$ s, write the mathematical sequence consisting of the samples taken from the interval $0 \le t \le 1$.
   (b) Repeat the previous part for the sampling period $T_s = \frac{1}{16}$ s.
   (c) Which sampling period is preferred $T_s = \frac{1}{4}$ s or $T_s = \frac{1}{16}$ s?

Fig. 1.107 Continuous time signal

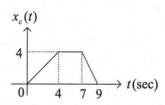

Fig. 1.108 Two lines for
Question-6

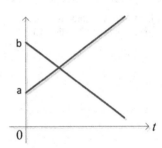

Fig. 1.109 Fourier transform
of a continuous time signal

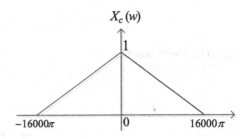

Fig. 1.110 Fourier transform
of a continuous time signal

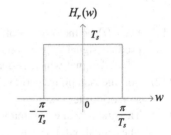

(8) The continuous time signal $x_c(t)$ is sampled with sampling period $T_s = \frac{1}{5000}$ s
     and the digital signal $x[n] = x_c(nT_s)$ is obtained. The Fourier transform of the
     continuous time signal is depicted in Fig. 1.109. Draw the Fourier transform
     of the digital signal $x[n]$.

(9) If $T_s = \frac{1}{8}$ s and $x[n] = [2 \quad -3 \quad 5 \quad 1 \quad 2 \quad 3 \quad 1.5 \quad 4.3 \quad 2.5 \quad -2.5 \quad 2]$,
     then draw the graph of

$$x_s(t) = \sum_{n=-\infty}^{\infty} x[n]\delta(n - T_s).$$

(10) The Fourier transform of a continuous time signal is depicted in Fig. 1.110.
     Using inverse Fourier transform formula, calculate the time domain expression
     of this signal.

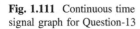

Fig. 1.111 Continuous time signal graph for Question-13

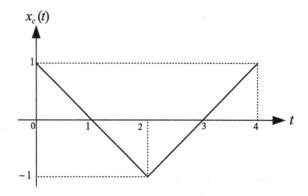

Fig. 1.112 Fourier transform of a continuous time signal

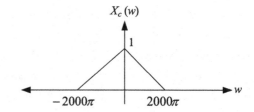

(11) Let $x_s(t)$ be the product of $x_c(t)$ and the impulse train function $s(t)$. Using the product signal expression, write the mathematical expression for the reconstructed signal which is evaluated as $x_r(t) = x_s(t) * h_r(t)$.

(12) For the sampling period $T_s = \frac{1}{8}$ s, draw the linearly approximated reconstruction filter graph.

(13) The graph of the continuous time signal $x_c(t)$ is displayed in Fig. 1.111. The signal $x_c(t)$ is sampled with sampling periods $T_s = 1$ s, $T_s = \frac{1}{4}$ s and $T_s = \frac{1}{8}$ s. Find the digital signal $x[n]$ for each sampling period.

(14) A continuous time signal is sampled with sampling period $T_s = \frac{1}{8}$ s and the digital signal $x[n] = \begin{bmatrix} 1 & 0.7 & 0 & -0.7 & -1 & -0.7 & 0 & 0.7 \end{bmatrix}$ is obtained. Using the approximated triangle reconstruction filter, rebuild the continuous time signal.

(15) The Fourier transform of a continuous time signal $x_c(t)$ is depicted in Fig. 1.112. The continuous time signal is sampled with sampling period $T_s = \frac{1}{3000}$ s and the digital signal $x[n] = x_c(t)|_{t=nT_s}$ is obtained. Draw the Fourier transform of $x[n]$.

(16) The continuous time signal $x_c(t) = \cos(2\pi \times 100 \times t) + \cos(2\pi \times 400 \times t)$ is sampled with sampling frequency f_s. How should f_s be chosen such that no aliasing occurs in the spectrum of digital signal.

(17) A continuous time signal is sampled with sampling frequency $f_s = 1000$ Hz. How many samples per second are taken from continuous time signal?

Chapter 2
Multirate Signal Processing

Digital signals are obtained from continuous time signals via sampling operation. Continuous time signals can be considered as digital signals having infinite number of samples. Sampling is nothing but selecting some of these samples and forming a mathematical sequence called digital signal. And these digital signals can be in periodic or non-periodic forms. The number of samples taken from a continuous time signal per-second is determined by sampling frequency. As the sampling frequency increases, the number of samples taken from a continuous time signal per-second increases, as well. As the technology improves, new and better electronic devices are being produced. This also brings the compatibility problem between old and new devices. One such problem is the speed issue of the devices. Consider a communication device transmitting digital samples taken from a continuous time signal at a high speed. This means high sampling frequency, as well. If the speed of the receiver device is not as high as the speed of the transmitter device, then the receiver device cannot accommodate the samples taken from the transmitter. This results in communication error. Hence, we should be able to change the sampling frequency according to our needs.

We should be able to increase or decrease the sampling frequency without changing the hardware. We can do this using additional hardware components at the output of the sampling devices. One way of decreasing the sampling frequency is the elimination of some of the samples of the digital signal. This is also called sampling of digital signals, or decimation of digital signals, or compression of digital signals. On the other hand, after digital transmission, at the receiver side before digital to analog conversion operation, we can increase the number of samples. This is called upsampling, or increasing sampling rate, or increasing sampling frequency. If we have more samples for a continuous time signal, when it is reconstructed from its samples, we obtain a better continuous time signal. In this chapter, we will learn how to manipulate digital signals, which means, changing their sampling rates, reconstruction of a long digital sequence from a short version of it, de-multiplexing and multiplexing of digital signals via hardware units etc.

© Springer Nature Singapore Pte Ltd. 2018
O. Gazi, *Understanding Digital Signal Processing*, Springer Topics
in Signal Processing 13, DOI 10.1007/978-981-10-4962-0_2

2.1 Sampling Rate Reduction by an Integer Factor (Downsampling, Compression)

To represent a continuous time by digital sequences, we take samples from the continuous time signal according a sampling frequency and form a mathematical sequence. If the mathematical sequence contains too many samples, we can omit some of these samples and keep the rest of the samples for transmission, storage, processing etc.

Let's give another example from real life. Assume that you want to send 500 students to a university in a foreign country. The selected students represent your university and from each department 10 students were selected. Later on you think that the travel cost of 500 students is too much and decide on reducing the number of selected students.

A continuous time signal can be considered as a digital signal containing infinite number of samples for any time interval. Sampling of analog signals is nothing but selecting a finite number of samples from the infinite sample sets of the analog signals for the given time interval. The downsampling operation can be considered as the sampling of digital signals. In this case a digital signal containing a number of samples for a given time interval is considered and for the given interval, some of the samples of the digital signal are selected and a new digital signal is formed. This operation is called downsampling. During the downsampling some of the samples of a digital signal are selected and the remaining samples are omitted.

The downsampling operation is illustrated in Fig. 2.1 where $x[n]$ is the signal to be downsampled and $y[n]$ is the signal obtained after downsampling $x[n]$, i.e., after omitting sampled from $x[n]$, and M is the downsampling factor.

Given $x[n]$ to find the compressed signal, i.e., downsampled signal, $y[n]$, we divide the time axis of $x[n]$ by M and keep only integer division results and omit all non-integer division results. Let's illustrate this operation by an example.

Example 2.1 A digital signal expressed as a mathematical sequence is given as

$$x[n] = [3.3 \quad -2.5 \quad 1.2 \quad 4.5 \quad 5.5 \quad -2.3 \quad \underset{n=0}{\underbrace{5.0}} \quad 6.2 \quad 3.4 \quad 2.3 \quad -4.4 \quad 3.2 \quad 2.0]$$

Find the downsampled $y[n] = x[3n]$.

Fig. 2.1 Downsampling operation

$$x[n] \longrightarrow \boxed{\downarrow M} \longrightarrow y[n] = x[Mn]$$

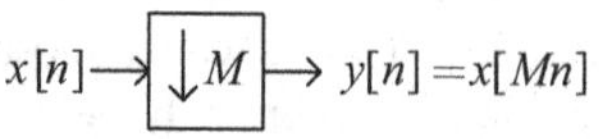

Solution 2.1 Let's write the time index values of the signal, $x[n]$ explicitly follows

$$x[n] = [\underbrace{3.3}_{n=-6} \quad \underbrace{-2.5}_{n=-5} \quad \underbrace{-1.2}_{n=-4} \quad \underbrace{4.5}_{n=-3} \quad \underbrace{5.5}_{n=-2} \quad \underbrace{-2.3}_{n=-1} \quad \underbrace{5.0}_{n=-0} \quad \underbrace{6.2}_{n=1} \quad \underbrace{3.4}_{n=2}$$

$$\underbrace{2.3}_{n=3} \quad \underbrace{-4.4}_{n=4} \quad \underbrace{3.2}_{n=5} \quad \underbrace{2.0}_{n=6}].$$

In the second step, we divide the time axis of $x[n]$ by 3, this is illustrated in

$$[\underbrace{\mathbf{3.3}}_{\mathbf{n=-\frac{6}{3}}} \quad \underbrace{-2.5}_{n=-\frac{5}{3}} \quad \underbrace{-1.2}_{n=-\frac{4}{3}} \quad \underbrace{\mathbf{4.5}}_{\mathbf{n=-\frac{3}{3}}} \quad \underbrace{5.5}_{n=-\frac{2}{3}} \quad \underbrace{-2.3}_{n=-\frac{1}{3}} \quad \underbrace{\mathbf{5.0}}_{\mathbf{n=-\frac{0}{3}}} \quad \underbrace{6.2}_{n=\frac{1}{3}} \quad \underbrace{3.4}_{n=\frac{2}{3}} \quad \underbrace{\mathbf{2.3}}_{\mathbf{n=\frac{3}{3}}}$$

$$\underbrace{-4.4}_{n=\frac{4}{3}} \quad \underbrace{3.2}_{n=\frac{5}{3}} \quad \underbrace{\mathbf{2.0}}_{\mathbf{n=\frac{6}{3}}}].$$

where divisions' yielding integer results are shown in bold numbers and these divisions are given alone as follows

$$[\underbrace{3.3}_{n=-\frac{6}{3}} \quad \underbrace{4.5}_{n=-\frac{3}{3}} \quad \underbrace{5.0}_{n=-\frac{0}{3}} \quad \underbrace{2.3}_{n=-\frac{3}{3}} \quad \underbrace{2.0}_{n=\frac{6}{3}}]$$

and when the divisions are done, we obtain the downsampled signal as

$$y[n] = [\underbrace{3.3}_{n=-2} \quad \underbrace{4.5}_{n=-1} \quad \underbrace{5.0}_{n=0} \quad \underbrace{2.3}_{n=1} \quad \underbrace{2.0}_{n=1}]$$

As it is seen from the previous example, downsampling a digital signal by M means that from every M samples of the digital signal only one of them is selected and the rest of them are eliminated. As an example, if $y[n] = x[6n]$, then from every 6 samples of $x[n]$ only one of them is kept and the other 5 samples are omitted.

Now we ask the question, if sampling frequency is f_s and downsampling factor is M, after downsampling operation how many samples per-second are available at the downsampler output? The answer is given in the block diagram in Fig. 2.2.

Where $\lceil . \rceil$ is the upper-floor operation. If f_s is a multiple of M, the diagram in Fig. 2.2 reduces to the one in Fig. 2.3.

Example 2.2 Interpret the block diagram given in Fig. 2.4.

Solution 2.2 At the input of the block, we receive 300 samples per-second which are obtained from an analog signal via sampling operation. At the output of the downsampler only 1 of every 3 samples is kept and the other 2 samples are omitted.

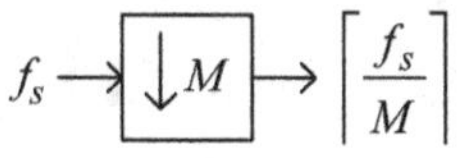

Fig. 2.2 Sampling frequency
at the downsampler output

$$f_s \rightarrow \boxed{\downarrow M} \rightarrow \left\lceil \frac{f_s}{M} \right\rceil$$

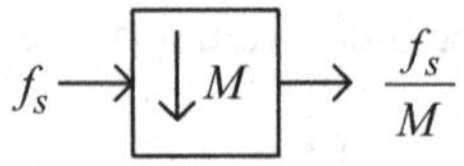

Fig. 2.3 Sampling frequency at the downsampler output when f_s is a multiple of M

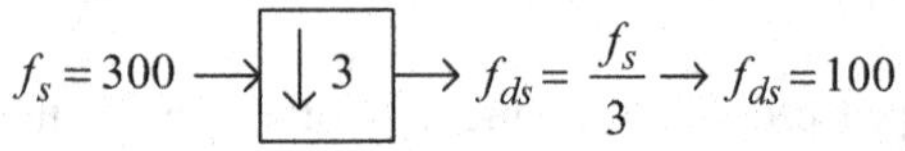

Fig. 2.4 Downsampler for Example 2.2

That means at the output of the downsampler, 100 samples every per-second are released.

Example 2.3 Find the Fourier series representation of

$$p[n] = \sum_{r=-\infty}^{\infty} \delta[n - rM]. \tag{2.1}$$

Solution 2.3 The given signal is a periodic signal with period M. Its Fourier series coefficients are computed as

$$P[k] = \frac{1}{M} \sum_{n=-\frac{M-1}{2}}^{\frac{M+1}{2}} p[n] e^{-j\frac{2\pi}{M}kn} \rightarrow P[k] = \frac{1}{M} \sum_{n=-\frac{M-1}{2}}^{\frac{M+1}{2}} \delta[n] e^{-j\frac{2\pi}{M}kn} \rightarrow P[k] = \frac{1}{M}. \tag{2.2}$$

Using the Fourier series coefficients in (2.2), the Fourier series representation of (2.1) can be written as

$$p[n] = \sum_{k,M} P[k] e^{j\frac{2\pi}{M}kn} \rightarrow p[n] = \frac{1}{M} \sum_{k,M} e^{j\frac{2\pi}{M}kn}. \tag{2.3}$$

The mathematical expression $p[n] = \sum_{r=-\infty}^{\infty} \delta[n - rM]$ can also be written as

$$p[n] = \begin{cases} 1 & if \quad n = 0, \pm M, \pm 2M, \ldots \\ 0 & otherwise. \end{cases} \tag{2.4}$$

And equating the right hand sides of (2.3) and (2.4) to each other, we get the equality

$$\frac{1}{M} \sum_{k=0}^{M-1} e^{j\frac{2\pi}{M}kn} = \begin{cases} 1 & n = 0, \pm M, \pm 2M, \ldots \\ 0 & otherwise. \end{cases} \tag{2.5}$$

For the expression in (2.5), if we change the sign of n appearing on both sides of the equation, we obtain an alternative expression for (2.5) as

$$\frac{1}{M}\sum_{k=0}^{M-1}e^{-j\frac{2\pi}{M}kn} = \begin{cases} 1 & n = 0, \pm M, \pm 2M, \ldots \\ 0 & otherwise. \end{cases} \tag{2.6}$$

2.1.1 Fourier Transform of the Downsampled Signal

Let's find the Fourier transform of the compressed signal $y[n] = x[Mn]$. The Fourier transform of $y[n]$ can be calculated using

$$Y_n(w) = \sum_{n=-\infty}^{\infty} x[Mn]e^{-jwn} \tag{2.7}$$

where defining $r \triangleq Mn$, we obtain

$$Y_n(w) = \sum_{r=0,\pm M,\pm 2M} x[r]e^{-jw\frac{r}{M}} \tag{2.8}$$

which can be written after parameter changes as

$$Y_n(w) = \sum_{n=0,\pm M,\pm 2M} x[n]e^{-jw\frac{n}{M}} \tag{2.9}$$

The frontiers of the sum symbol in (2.9) can be changed to $-\infty$ and ∞ if (2.1) is used in (2.9) as

$$Y_n(w) = \sum_{n=-\infty}^{\infty} x[n] \sum_{r=-\infty}^{\infty} \delta[n - rM]e^{-jw\frac{n}{M}}$$

where replacing $\sum_{r=-\infty}^{\infty} \delta[n - rM]$ by its Fourier series representation, we get

$$Y_n(w) = \sum_{n=-\infty}^{\infty} x[n]\frac{1}{M}\sum_{k,M} e^{-j\frac{2\pi}{M}kn}e^{-jw\frac{n}{M}} \tag{2.10}$$

which can be rearranged as

$$Y_n(w) = \frac{1}{M}\sum_{k,M}\underbrace{\sum_{n=-\infty}^{\infty} x[n]e^{-j\frac{w+k2\pi}{M}n}}_{=X_n\left(\frac{w+k2\pi}{M}\right)} \tag{2.11}$$

The expression in (2.11) can be reduced to

$$Y_n(w) = \frac{1}{M} \sum_{k,M} X_n \left(\frac{w + k2\pi}{M} \right). \tag{2.12}$$

In (2.10), if (2.5) was used, then we would obtain

$$Y_n(w) = \frac{1}{M} \sum_{k,M} X_n \left(\frac{w - k2\pi}{M} \right). \tag{2.13}$$

Hence, considering (2.12) and (2.13), we can write the Fourier transform of $y[n] = x[Mn]$ as

$$Y_n(w) = \frac{1}{M} \sum_{k=0}^{M-1} X_n \left(\frac{w \pm k2\pi}{M} \right). \tag{2.14}$$

Example 2.4 If $y[n] = x[Mn]$ the relation between Fourier transforms of $x[n]$ and $y[n]$ is given as

$$Y_n(w) = \frac{1}{M} \sum_{k=0}^{M-1} X_n \left(\frac{w \pm k2\pi}{M} \right).$$

Using the inverse Fourier transform expression for $y[n]$, i.e.,

$$y[n] = \frac{1}{2\pi} \int_{w=0}^{2\pi} Y_n(w) e^{jwn} dw \tag{2.15}$$

show that $y[n] = x[Mn]$.

Solution 2.4 The inverse Fourier transform is given as

$$y[n] = \frac{1}{2\pi} \int_{0}^{2\pi} Y_n(w) e^{jwn} dw$$

where inserting

$$Y_n(w) = \frac{1}{M} \sum_{k=0}^{M-1} X_n \left(\frac{w \pm k2\pi}{M} \right)$$

we get

$$y[n] = \frac{1}{2\pi M} \sum_{k=0}^{M-1} \int_0^{2\pi} X_n\left(\frac{w+k2\pi}{M}\right) e^{jwn} dw \qquad (2.16)$$

In (2.16), if we let $\lambda = \frac{w+k2\pi}{M}$, then $dw = Md\lambda$, and changing the frontiers of the integral (2.16) reduces to

$$y[n] = \frac{1}{2\pi} \sum_{k=0}^{M-1} \int_{\frac{k2\pi}{M}}^{\frac{(k+1)2\pi}{M}} X_n(\lambda) e^{jM\lambda n} d\lambda. \qquad (2.17)$$

If (2.17) is expanded for all k values, we obtain

$$y[n] = \frac{1}{2\pi} \int_0^{\frac{2\pi}{M}} X_n(\lambda) e^{jM\lambda n} d\lambda + \frac{1}{2\pi} \int_{\frac{2\pi}{M}}^{\frac{4\pi}{M}} X_n(\lambda) e^{jM\lambda n} d\lambda + \cdots$$

$$+ \frac{1}{2\pi} \int_{\frac{M-1}{M}2\pi}^{2\pi} X_n(\lambda) e^{jM\lambda n} d\lambda \qquad (2.18)$$

where using the property $\int_a^b (\cdot) + \int_b^c (\cdot) = \int_a^c (\cdot)$ and changing λ with w, we get the expression

$$y[n] = \frac{1}{2\pi} \int_0^{2\pi} X_n(w) e^{jMwn} dw. \qquad (2.19)$$

When (2.19) is compared to

$$x[n] = \frac{1}{2\pi} \int_0^{2\pi} X_n(w) e^{jwn} dw$$

it is seen that $y[n] = x[Mn]$.

2.1.2 How to Draw the Frequency Response of Downsampled Signal

To draw the graph of

$$Y_n(w) = \frac{1}{M} \sum_{k=0}^{M-1} X_n\left(\frac{w - k2\pi}{M}\right)$$

students usually expand the summation as

$$Y_n(w) = \frac{1}{M} X_n\left(\frac{w}{M}\right) + \frac{1}{M} X_n\left(\frac{w - 2\pi}{M}\right) + \frac{1}{M} X_n\left(\frac{w - 4\pi}{M}\right) + \cdots \quad (2.20)$$

and try to draw each shifted graph and sum the shifted graphs. However, this approach is too time consuming and error-prone. Instead of this approach, we will suggest a simpler method to draw the graph of $Y_n(w)$ as explained in the following lines.

Since $Y_n(w)$ is the Fourier transform of the digital signal $y[n]$, then $Y_n(w)$ is a periodic signal and its period equals to 2π.

To draw the graph of $Y_n(w)$, we can follow the following steps.

Step 1: First one period of $X_n(w)$ around origin is drawn. For this purpose, the frequency interval is chosen as $-\pi < w \le \pi$.

Step 2: Considering one period of $X_n(w)$ around origin, we draw one period of $\frac{1}{M} X_n\left(\frac{w}{M}\right)$. To draw (in one period) the graph of $\frac{1}{M} X_n\left(\frac{w}{M}\right)$, we multiply the horizontal axis of $X_n(w)$ by M, and multiply the vertical axis of $X_n(w)$ by $\frac{1}{M}$.

Step 3: In Step 3, we shift the resulting graph in Step 2 to the left and right by multiples of 2π and sum the shifted replicas.

Let's now give an example to illustrate the topic.

Example 2.5 One period of the Fourier transform of $x[n]$ is depicted in Fig. 2.5. Draw the Fourier transform of $y[n] = x[2n]$, i.e., draw $Y_n(w)$.

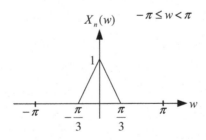

Fig. 2.5 One period of the Fourier transform of $x[n]$

Solution 2.5 First let's draw the graph of $Y_{1n}(w) = \frac{1}{2}X_n(\frac{w}{2})$. For this purpose, we multiply the frequency axis of $X_n(w)$ by 2 and vertical axis of $X_n(w)$ by $\frac{1}{2}$. The resulting graph is shown in Fig. 2.6.

In the second step, we shift the graph of $Y_{1n}(w)$ to the left and right by multiples of 2π and sum the shifted graphs. In other words, we draw the graph of $Y_n(w) = \sum_{k=-\infty}^{\infty} Y_{1n}(w - k2\pi)$. The shifted graphs and their summation result are depicted in Figs. 2.7, 2.8, and 2.9.

Right Shifted Functions:

Left Shifted Functions:

Sum of the Shifted Functions:

Exercise: One period of the Fourier transform of $x[n]$ is depicted in Fig. 2.10. Draw the Fourier transform of $y[n] = x[3n]$, i.e., draw $Y_n(w)$.

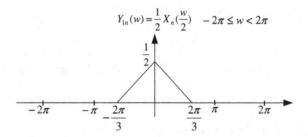

Fig. 2.6 The graph of $\frac{1}{2}X_n(\frac{w}{2})$

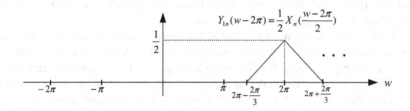

Fig. 2.7 Right shifted functions

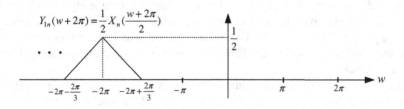

Fig. 2.8 Left shifted functions

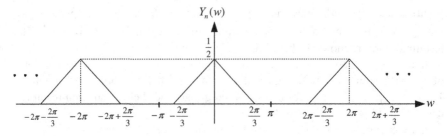

Fig. 2.9 Sum of the shifted functions

Fig. 2.10 One period of the
Fourier transform of $x[n]$

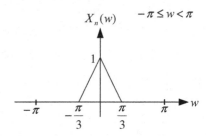

2.1.3 Aliasing in Downsampling

A digital signal is nothing but a mathematical sequence obtained via sampling of a continuous time signal. If we have sufficient number of samples, we can reconstruct the continuous time signal from its samples.

If we have too many samples, generated during the sampling operation we can eliminate some of these excessive samples via the downsampling operation. However, while performing the downsampling operation, we should be careful to keep sufficient number of samples in the digital signal such that the reconstruction of the continuous time signal is still possible after downsampling operation.

If we eliminate a number of samples more than a threshold value, the rest of the samples may not be sufficient to reconstruct the continuous time signal and this effect is seen as the aliasing in the spectrum graph of the downsampled signal.

Example 2.6 Assume that we have a low pass continuous time signal with bandwidth $f_N = 40$ Hz. We choose the sampling frequency according to the criteria $f_s > 2f_N \rightarrow f_s > 80$ as $f_s = 120$. This means that we take 120 samples per-second from the continuous time signal. However, our chosen sampling frequency is not very cost efficient.

The lower limit for the sampling frequency is $f_s > 80$ which means that the minimum sampling frequency can be chosen as $f_s = 81$. However we use $f_s = 120$ which means that every per-second we transmit $120 - 81 = 39$ excessive samples which are not necessary to reconstruct the continuous time signal. We can reconstruct the continuous time signal using only 81 samples. We can omit the excessive 39 samples via downsampling operation.

Let's now determine the criteria for no aliasing in downsampling operation. After downsampling operation, we have $\frac{f_s}{M}$ remaining samples per-second. If this number of remaining samples is greater than $2f_N$, then no aliasing occurs. That is if

$$\frac{f_s}{M} > 2f_N \rightarrow M < \frac{f_s}{2f_N} \tag{2.21}$$

is satisfied, then aliasing is not seen in the spectrum of the downsampled signal. Let's simplify (2.21) more as

$$M < \frac{f_s}{2f_N} \rightarrow M < \frac{1}{2\underbrace{T_s f_N}_{f_D}} \tag{2.22}$$

where f_D is the digital frequency, and manipulating more, we have

$$M < \frac{1}{2f_D} \rightarrow M < \frac{\pi}{2\pi f_D} \rightarrow M < \frac{\pi}{w_D} \rightarrow Mw_D < \pi \tag{2.23}$$

where w_D is the angular digital frequency.

Let's now graphically illustrate the no aliasing criteria after downsampling operation. Assume that one period of the Fourier transform of the digital signal $x[n]$ to be downsampled is given as in Fig. 2.11. Let $y[n] = x[Mn]$ be the downsampled signal.

Depending on the value of M, we can draw the two possible graphs of $\frac{1}{M}X_n\left(\frac{w}{M}\right)$ as shown in Figs. 2.12 and 2.13.

When the graph in Fig. 2.12 is shifted to the left and right by multiples of 2π, no overlapping occurs among shifted graphs. However, this case does not hold for the graph shown in Fig. 2.13. If the graph shown in Fig. 2.13 is shifted to the left and right by multiples of 2π, overlapping is observed between shifted replicas, and this situation is depicted in Fig. 2.14.

Example 2.7 The continuous time signal $x_c(t) = \cos(6000\pi t)$ is sampled with sampling period $T_s = \frac{1}{8000}$ and the digital sequence $x[n]$ is obtained. Next the digital

Fig. 2.11 One period of the Fourier transform of the digital signal $x[n]$ to be downsampled

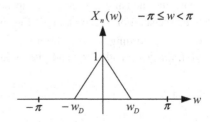

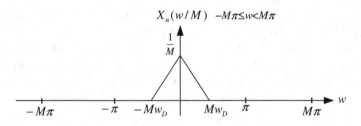

Fig. 2.12 Case-1: Graph of $\frac{1}{M}X_n\left(\frac{w}{M}\right)$

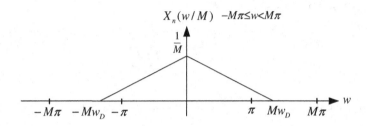

Fig. 2.13 Case-2: Graph of $\frac{1}{M}X_n\left(\frac{w}{M}\right)$

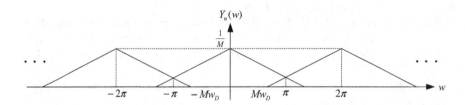

Fig. 2.14 Aliasing in downsampled signal spectrum graph

signal $x[n]$ is downsampled and $y[n] = x[4n]$ is obtained. Decide whether aliasing occurs in spectrum of $y[n]$ or not.

Solution 2.7 If the given continuous time signal is compared to $\cos(2\pi ft)$, the frequency of the continuous time signal is found as $f = 3000$ Hz. And the sampling frequency is $f_s = 8000$ Hz. After downsampling operation sampling frequency reduces to $f_s = \frac{8000}{4} = 2000$ Hz and this value is less than $2f = 6000$ Hz. This means that aliasing is seen in the spectrum of $y[n]$.

Exercise: For the system in Fig. 2.15, $x_c(t) = \cos(5000\pi t)$, $T_s = \frac{1}{10,000}$, and $M = 2$. According to given information, draw the Fourier transforms of the signals $x_c(t), x[n], y[n]$, and $y_r(t)$, and also write the time domain expression for $y_r(t)$.

$$y[n] = x[Mn]$$

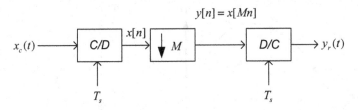

Fig. 2.15 Signal processing system for exercise

2.1.4 Interpretation of the Downsampling in Terms of the Sampling Period

If $x[n] = x_c(nT_s)$, then for the downsampled signal $y[n] = x[Mn] \rightarrow y[n] = x_c(n \underbrace{MT_s}_{T_s'})$ new sampling period is $T_s' = MT_s$ which is an integer multiple of T_s. The digital signal obtained from $x_c(t)$ using sampling period T_s is shown in Fig. 2.16. The digital signal $x[n]$ in Fig. 2.16 is written as a mathematical sequence as

$$x[n] = [\cdots a \quad b \quad c \quad d \quad e \quad f \quad \underbrace{g}_{n=0} \quad h \quad i \quad j \quad k \quad l \quad m \cdots].$$

Now consider $y[n] = x[2n] \rightarrow y[n] = x_c(n2T_s)$, in this case the samples are taken from $x_c(t)$ at every $T_s' = 2T_s$. This operation is illustrated in Fig. 2.17.

The digital signal $y[n]$ in Fig. 2.17 can be written as a mathematical sequence as

$$y[n] = [\cdots a \quad c \quad e \quad \underbrace{g}_{n=0} \quad i \quad k \quad m \cdots].$$

Similarly, if $g[n] = x[4n] \rightarrow g[n] = x_c(n4T_s)$, the samples are taken from $x_c(t)$ at every $T_s' = 4T_s$. This operation is illustrated in Fig. 2.18.

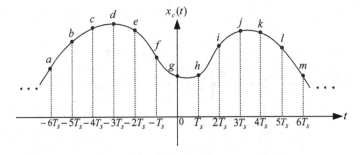

Fig. 2.16 Sampling of the continuous time signal with sampling period T_s

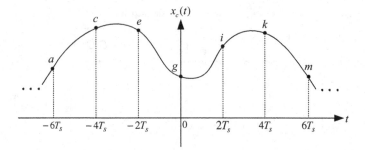

Fig. 2.17 Sampling of the continuous time signal with sampling period $2T_s$

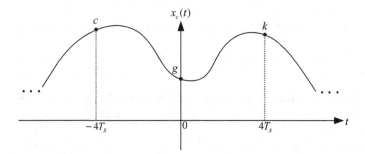

Fig. 2.18 Sampling of the continuous time signal with sampling period $4T_s$

The digital signal $g[n]$ in Fig. 2.18 can be written as a mathematical sequence as

$$y[n] = [\cdots c \quad \underbrace{g}_{n=0} \quad k \cdots].$$

Example 2.8 For the signal processing system given in Fig. 2.19, $x_c(t) = \cos(5000\pi t)$, $T_s = \frac{1}{8000}$, and $M = 3$. Using the given information, calculate and draw the Fourier transforms of the signals $x_c(t), x[n], y[n]$, and $y_r(t)$. Besides, write the time domain expression for $y_r(t)$.

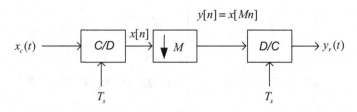

Fig. 2.19 Signal processing system for Example 2.8

Solution 2.8 Before starting to the solution, let's provide some background information as

$$\mathrm{Cos}(\theta) = \frac{1}{2}\left(e^{j\theta} + e^{-j\theta}\right) \quad FT\{e^{jw_0 t}\} = 2\pi\delta(w - w_0) \tag{2.24}$$

$$FT\{\cos(w_N t)\} = \pi(\delta(w - w_N) + \delta(w + w_N)). \tag{2.25}$$

Accordingly, the Fourier transform of $x_c(t)$ is found as

$$X_c(w) = \pi(\delta(w - 5000\pi) + \delta(w + 5000\pi)).$$

and graphically it is shown in Fig. 2.20.

For the given example, since $f_s > 2f_N \rightarrow 8000 > 2 \times 2500$ criteria is satisfied, no aliasing is observed in the Fourier transform of $x[n]$, and for this reason, one period of the Fourier transform of $x[n]$ for the interval $-\pi \le w < \pi$ equals $X_n(w) = \frac{1}{T_s} X_c\left(\frac{w}{T_s}\right)$ which is depicted in Fig. 2.21.

For the downsampled signal, we have $y[n] = x[3n]$, let's draw one period of $Y_n(w) = \frac{1}{3} X_n\left(\frac{w}{3}\right)$ using one period of $X_n(w)$ around origin as in Fig. 2.22 where impulses are labeled with letters so that we can distinguish them while forming the Fourier transform of $y[n]$.

If the graph in Fig. 2.22 is carefully inspected, we see that after downsampling operation one period of the Fourier transform of the downsampled signal extends beyond the interval $(-\pi, \pi)$ in frequency axis. This means that the number of samples omitted is greater than the allowed threshold and for this reason perfect reconstruction of the continuous time signal is not possible anymore. It may be reconstructed with some distortion or the reconstructed signal may be a totally

Fig. 2.20 Fourier transform of $x_c(t)$ in Example 2.8

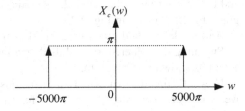

Fig. 2.21 One period of the Fourier transform of $x[n]$ for Example 2.8

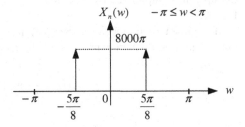

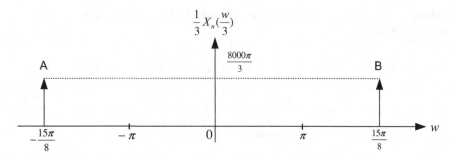

Fig. 2.22 The graph of $\frac{1}{3}X_n\left(\frac{w}{3}\right)$ for Example 2.8

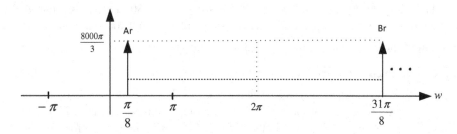

Fig. 2.23 One period of $Y_n(w)$ shifted to the right by 2π

different one. The amount of distortion in the reconstructed continuous time signal depends on the rate of the omitted samples, i.e., rate of the compression or rate of the downsampling. As the number of omitted samples increases, the amount of distortion in the reconstructed signal increases, as well.

To get the graph of $Y_n(w)$, we shift its one period depicted in Fig. 2.22 to the left and to the right by multiples of 2π and sum the shifted replicas. The right shifted graph by 2π is given in Fig. 2.23.

And the left shifted graph by 2π is shown in Fig. 2.24a.

Summing the centered, right shifted, and left shifted graphs, we get the graph of $Y_n(w)$ as shown in Fig. 2.24b.

Now let's find the expression for the reconstructed signal $y_r(t)$. For this purpose, we consider the graph of $Y_n(w)$ for the interval $-\pi \leq w < \pi$ and draw $Y_r(w) = T_s X_n(T_s w)$. To achieve this, we divide the frequency axis by T_s and multiply the amplitudes by T_s. These operations generate the graph depicted in Fig. 2.25.

If the inverse Fourier transform of $Y_r(w)$ depicted in Fig. 2.25 is calculated, we obtain the time domain expression of the reconstructed signal as

$$y_r(t) = \frac{1}{3}\cos\left(1000\pi t\right)$$

(a)

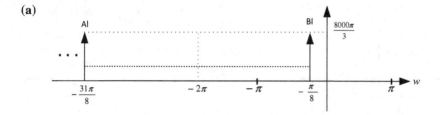

(b)

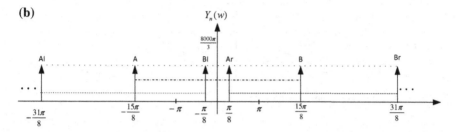

Fig. 2.24 **a** One period of $Y_n(w)$ shifted to the left by 2π. **b** The graph of $Y_n(w)$ for Example 2.8

Fig. 2.25 Fourier transform of the reconstructed signal for Example 2.8

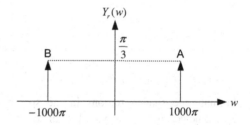

which is quite different from the sampled signal $x_c(t) = \cos(5000\pi t)$. The reason for this is that during the downsampling operation too many samples, beyond the allowable threshold, are omitted and this resulted in aliasing in frequency domain and perfect reconstruction of the original signal is not possible anymore.

Question: During the downsampling operation we have to omit more samples than the number of allowable one. However, we want to decrease the effect of aliasing at the spectrum of the digital signal. What can we do for this?

Answer: If $y[n] = x[Mn]$ alising occurs in $Y_n(w)$, if the largest frequency of $X_n(w)$ in the interval $-\pi \leq w < \pi$ is greater than $\frac{\pi}{M}$. This situation is depicted in Fig. 2.26.

For the conversion of $y[n]$ to continuous time signal $y_r(t)$, the portion of $Y_n(w)$ for the interval $-\pi \leq w < \pi$ in Fig. 2.26 is used. This portion is depicted alone in Fig. 2.27.

As it is seen from Fig. 2.27, the overlapping shaded parts cause distortion in the reconstructed signal. Then how can we decrease the distortion amount? If we can

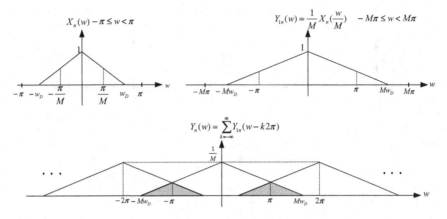

Fig. 2.26 Aliasing case in downsampled signal

Fig. 2.27 $Y_n(w)$, $-\pi \leq w < \pi$

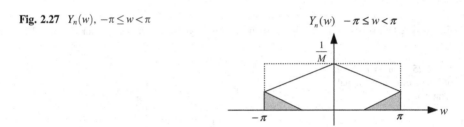

eliminate the shaded regions in the spectrum of the downsampled signal, the reconstructed signal will have less distortion.

However due to the clipping of the parts extending beyond the interval $(-\pi, \pi)$, some distortion will always be available in the reconstructed signal. This distortion is due to the information loss owing to the clipping of the spectrum regions in Fig. 2.26 for the intervals $\pi \leq w < Mw_d$ and $-M\pi \leq w < \pi$. What we do here is that we want try to decrease the amount of distortion, not complete elimination of it.

Then if we can get a spectrum graph for $Y_n(w)$, $-\pi \leq w < \pi$ as shown in Fig. 2.28 the reconstructed signal will have less distortion.

Fig. 2.28 After elimination of the overlapping shaded parts in Fig. 2.27

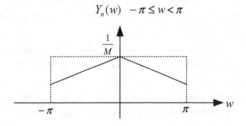

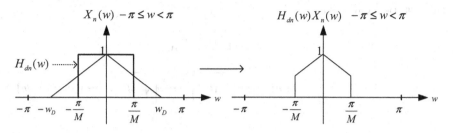

Fig. 2.29 Elimination of the high frequency parts by a decimator filter

We can omit the overlapping shaded parts if we can filter high frequency portions of $X_n(w)$ before downsampling operation, i.e., the portions of $X_n(w)$ for the intervals $\frac{\pi}{M} \leq w < \pi$ and $-\pi \leq w < -\frac{\pi}{M}$ should be filtered out. This can be achieved using a low pass filter as shown in bold lines Fig. 2.29. The lowpass filter clips the wigs of the signal that extends beyond the interval $(-\pi, \pi)$. And this clipping prevents the overlapping problem in downsampled signal spectrum.

The lowpass filter used in Fig. 2.29 is called decimator filter whose frequency domain expression for its one period around origin is written as

$$H_{dn}(w) = \begin{cases} 1 & if \quad |w| < \frac{\pi}{M} \\ 0 & if \quad \frac{\pi}{M} < |w| < \pi. \end{cases} \tag{2.26}$$

The time domain expression of the decimator filter can be computed using the inverse Fourier transform as

$$h_{dn}[n] = \frac{1}{2\pi} \int_{w,2\pi} H_{dn}(w)e^{jwn}dw \rightarrow h_{dn}[n] = \frac{1}{2\pi} \int_{-\frac{\pi}{M}}^{\frac{\pi}{M}} 1 \times e^{jwn}dw \tag{2.27}$$

yielding the expression

$$h_{dn}[n] = \frac{\sin\left(\frac{\pi n}{M}\right)}{\pi n} \rightarrow h_{dn}[n] = \frac{1}{M}\sin c\left(\frac{n}{M}\right). \tag{2.28}$$

The filtering process before downsampling operation is illustrated in Fig. 2.30. The system in Fig. 2.30 is called **decimator system**, and the overall operation in Fig. 2.30 is named as **decimation**.

For the system in Fig. 2.30, we have $Y_{1n}(w) = H_{dn}(w)X_n(w)$ and $y[n] = y_1[Mn]$. One period of $Y_n(w)$ is written as $Y_n(w) = \frac{1}{M}Y_{1n}\left(\frac{w}{M}\right)$, $-\pi \leq w < \pi$. One period of $Y_n(w)$ is shown in Fig. 2.31.

Fig. 2.30 Decimator system

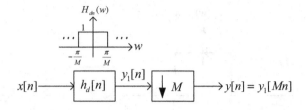

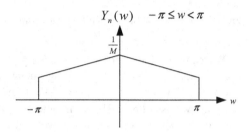

Fig. 2.31 One period of $Y_n(w)$

One period of $Y_n(w)$ can be expressed as

$$Y_{nop}(w) = \begin{cases} Y_n(w) & -\pi \leq w < \pi \\ 0 & otherwise \end{cases} \qquad (2.29)$$

which can be used for the calculation of the Fourier transform of $y[n]$ as

$$Y_n(w) = \sum_{k=-\infty}^{\infty} Y_{nop}(w - k2\pi). \qquad (2.30)$$

Considering Fig. 2.31 the graph of (2.30) can be drawn as in Fig. 2.32.

Exercise: If $y[n] = x[3n]$ and the Fourier transform of $x[n]$ for $-\pi \leq w < \pi$ is as given in Fig. 2.33, draw the Fourier transform of $y[n]$, i.e., draw $Y_n(w)$.

Downsampling can also be used for de-multiplexing operations, i.e., separating digital data to its components. We below give some examples to illustrate the use of downsampling for de-multiplexing operations.

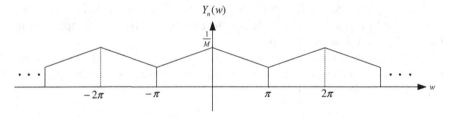

Fig. 2.32 Fourier transform of filtered and dowsampled signal

Fig. 2.33 One period of the Fourier transform of a digital signal

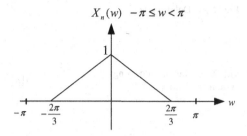

$X_n(w) \quad -\pi \le w < \pi$

Note: The simplest de-multiplexer is the serial to parallel converter.

Example 2.9 The delay system is described in Fig. 2.34.

If

$$x[n] = [1 \quad 2 \quad 3 \quad 4 \quad 5 \quad 6 \quad 7 \quad 8 \quad \underbrace{9}_{n=0} \quad 10 \quad 11 \quad 12 \quad 13 \quad 14 \quad 15]$$

find the output of each unit given in Fig. 2.35.

Solution 2.9 To get $y[n] = x[n - n_0], n_0 > 0$, it is sufficient to shift $n = 0$ pointer to the left by n_0 units in $x[n]$ sequence. For negative n_0, we shift the $n = 0$ pointer to the right by n_0 units. According to this information, $x[n - 1]$ can be calculated as

$$x[n - 1] = [1 \quad 2 \quad 3 \quad 4 \quad 5 \quad 6 \quad 7 \quad \underbrace{8}_{n=0} \quad 9 \quad 10 \quad 11 \quad 12 \quad 13 \quad 14 \quad 15].$$

If we divide the time axis by 2 and take only the integer division results, we get the signals

$$y_1[n] = [1 \quad 3 \quad 5 \quad 7 \quad \underbrace{9}_{n=0} \quad 11 \quad 13 \quad 15] \quad y_2[n] = [2 \quad 4 \quad 6 \quad \underbrace{8}_{n=0} \quad 10 \quad 12 \quad 14]$$

at the outputs of the downsamplers.

As it is seen from the obtained sequences, the system separates the odd and even indexed samples.

Fig. 2.34 Delay system

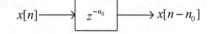

$x[n] \longrightarrow \boxed{z^{-n_0}} \longrightarrow x[n - n_0]$

Fig. 2.35 Signal processing system for Example 2.9

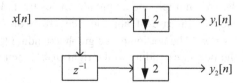

Fig. 2.36 Delay system

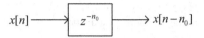

$$x[n] \longrightarrow \boxed{z^{-n_0}} \longrightarrow x[n-n_0]$$

Fig. 2.37 Signal processing
system for Example 2.10

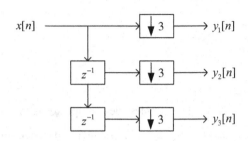

Example 2.10 The delay system is shown in Fig. 2.36.
If

$$x[n] = [1 \quad 2 \quad 3 \quad 4 \quad 5 \quad 6 \quad 7 \quad 8 \quad \underbrace{9}_{n=0} \quad 10 \quad 11 \quad 12 \quad 13 \quad 14 \quad 15]$$

find the output of each unit given in Fig. 2.37.

Solution 2.10 Following similar steps as in the previous example, we find the
digital signals at the outputs of the downsamplers as

$$y_1[n] = [3 \quad 6 \quad 9 \quad 12 \quad 15] \quad y_2[n] = [2 \quad 5 \quad 8 \quad 11 \quad 14]$$
$$y_3[n] = [1 \quad 4 \quad 7 \quad 10 \quad 13]$$

which are nothing but sub-sequences obtained by dividing data signal $x[n]$ into
non-overlapping sequences.

2.1.5 Drawing the Fourier Transform of Downsampled Signal in Case of Aliasing (Practical Method)

Let $y[n] = x[Mn]$ be the downsampled digital signal. To draw the Fourier transform
of $y[n]$ in case of aliasing, we follow the subsequent steps.

Step 1: First we draw the graph of $\frac{1}{M} X_n\left(\frac{w}{M}\right)$. For this purpose, we divide the
horizontal axis of the graph of $X_n(w)$ by $\frac{1}{M}$, i.e., we multiply the horizontal axis by
M, and multiply the amplitude values by $1/M$.
Step 2: In case of aliasing, the graph of $\frac{1}{M} X_n\left(\frac{w}{M}\right)$ extends beyond the interval
$(-\pi, \pi)$. The portion of the graph extending to the left of $-\pi$ is denoted by 'A', and
the potion extending to the right of π is denoted by 'B'.

Step 3: The portion of the graph denoted by 'A' in Step 2 is shifted to the right by 2π, and the portion denoted by 'B' is shifted to the left by 2π. The overlapping lines are summed and one period of $Y_n(w)$ around origin is obtained. Let's denote this one period by $Y_{n1}(w)$.

Step 4: In the last step, one period of $Y_n(w)$ around origin denoted by $Y_{n1}(w)$ is shifted to the left and right by multiples of 2π and all the shifted replicas are summed to get $Y_n(w)$, this is mathematically stated as

$$Y_n(w) = \sum_{k=-\infty}^{\infty} Y_{n1}(w - k2\pi).$$

Now let's explain these steps using graphics.

Let the Fourier transform of $x[n]$ be as shown in Fig. 2.38.

In case of aliasing, one period of $\frac{1}{M}X_n\left(\frac{w}{M}\right)$ around origin will be as shown in Fig. 2.39.

If Fig. 2.39 is inspected carefully, it is seen that the function $\frac{1}{M}X_n\left(\frac{w}{M}\right)$ takes values outside the interval $(-\pi, \pi)$ on horizontal axis. In Fig. 2.40, the shadowed triangles denoted by 'A' and 'B' show the portion of $\frac{1}{M}X_n\left(\frac{w}{M}\right)$ extending outside of $(-\pi, \pi)$.

Fig. 2.38 Fourier transform of $x[n]$

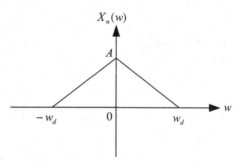

Fig. 2.39 One period of $\frac{1}{M}X_n\left(\frac{w}{M}\right)$ around origin in case of aliasing

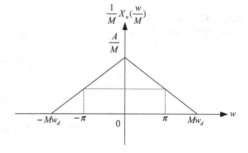

If the shadowed triangles 'A' and 'B' in Fig. 2.40 are shifted to the right and left by 2π, we obtain the graphic in Fig. 2.41. If the overlapping lines in Fig. 2.41 are summed, we obtain the graphic shown in bold lines in Fig. 2.42. As it is clear from Fig. 2.41, overlapping regions distorts the original signal. The amount of distortion depends on the widths of the shadowed triangles. In other words, as the function $\frac{1}{M}X_n\left(\frac{w}{M}\right)$ extends outside the interval $(-\pi, \pi)$ more, the amount of distortion on the original signal due to overlapping increases.

The graph obtained after summing the overlapping lines is depicted alone in Fig. 2.43.

Fig. 2.40 One period of $\frac{1}{M}X_n\left(\frac{w}{M}\right)$ around origin in case of aliasing

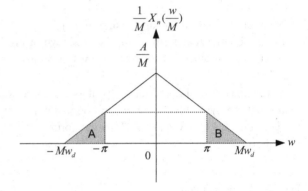

Fig. 2.41 Shaded parts shifted

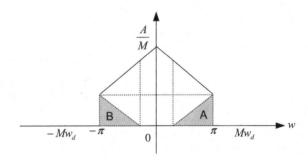

Fig. 2.42 Sum of the overlapping lines

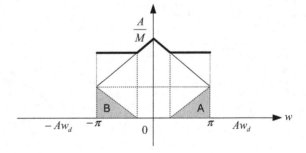

Fig. 2.43 The resulting graph after summing the overlapping lines

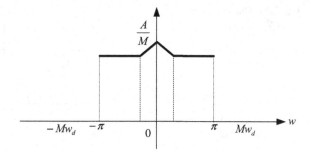

Fig. 2.44 One period of $X_n(w)$

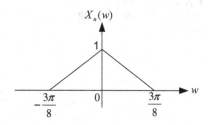

Exercise 2.11 The Fourier transform of $x[n]$, i.e., $X_n(w)$, is shown in Fig. 2.44. Draw the Fourier transform of the downsampled signal $y[n] = x[Mn]$, $M = 4$.

Solution 2.11

Step 1: First we draw the graph of $\frac{1}{M} X_n\left(\frac{w}{M}\right)$ as in Fig. 2.45.

For the graph of Fig. 2.45, the parts that fall outside of the interval $(-\pi, \pi)$ are denoted by the shaded triangles 'A' and 'B' in Fig. 2.46.

If the shaded parts 'A' and 'B' in Fig. 2.46 are shifted to the right and to the left by 2π, we obtain the graph in Fig. 2.47.

The equations of the overlapping line on the interval $(-\pi, -\pi/2)$ in Fig. 2.47 can be written as $\frac{1}{12\pi} w + \frac{1}{4}$ and $-\frac{1}{12\pi} w - \frac{1}{24}$, and when these equations are summed, we obtain $\frac{5}{24}$. In a similar manner, the sum of the equations of the overlapping line on the interval $(\pi/2, \pi)$ can be found as $\frac{5}{24}$. Hence one period of $Y_n(w)$ around origin can be drawn as shown in Fig. 2.48.

In the last step, shifting one period of $Y_n(w)$ to the left and right by multiples of 2π and summing the shifted replicas we obtain the graph of $Y_n(w)$.

Fig. 2.45 One period of $\frac{1}{M} X_n\left(\frac{w}{M}\right)$

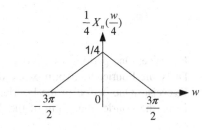

Fig. 2.46 One period of
$\frac{1}{M}X_n\left(\frac{w}{M}\right)$

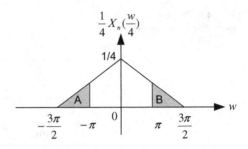

Fig. 2.47 *Shaded parts shifted to the right and to the left by 2π*

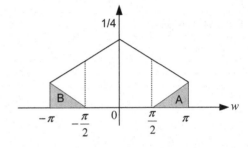

Fig. 2.48 One period of $Y_n(w)$ around origin

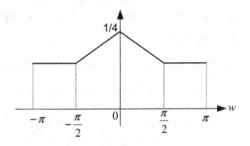

Fig. 2.49 One period of the Fourier transform of $x[n]$

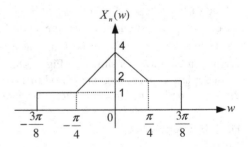

Exercise: One period of the Fourier transform of $x[n]$ is shown in Fig. 2.49. Draw the Fourier transform of the downsampled signal $y[n] = x[4n]$.

Exercise: One period of the Fourier transform of $x[n]$ is shown in Fig. 2.50. Draw the Fourier transform of the downsampled signal $y[n] = x[8n]$.

Fig. 2.50 $X_n(w)$ bir periyodu

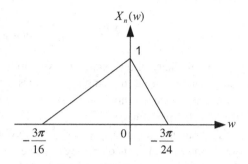

2.2 Upsampling: Increasing the Sampling Rate by an Integer Factor

Assume that we want to transmit an analog signal. For this purpose, we first take some samples from the continuous time signal and form a mathematical sequence, and this process is called sampling. To decrease the transmission overhead, we omit some of the digital samples and this process is called downsampling. After downsampling operation, we transmit the remaining samples. At the receiver side, for better reconstruction of the analog signal, we try to find a method to increase the number of digital samples. For this purpose, we try to find the samples omitted during the downsampling operation. After finding the omitted samples, we can reconstruct the analog signal in a better manner.

This means that first we reconstruct the original digital signal from downsampled digital signal then by using the reconstructed digital signal, we reconstruct the continuous time signal.

Reconstruction of the original digital signal from the downsampled signal includes a two-step process. The first step is called up sampling also named as signal-expansion. In this step, the compressed signal, i.e., downsampled signal, is expanded in time axis, and for the new time instants, 0 values are assigned for the new amplitudes. The second step is called interpolation which is the reconstruction part for the omitted digital samples. In this part, the 0 values assigned to new time amplitudes for the expanded signal are replaced by the estimated values.

Now let's explain the upsampling operation.

2.2.1 Upsampling (Expansion)

The block diagram of the upsampler (expander) is shown in Fig. 2.51.

The mathematical expression of the upsampling operation is

Fig. 2.51 Upsampling operation

$$y[n] = \begin{cases} x\left[\frac{n}{L}\right] & n = 0, \pm L, \pm 2L, \ldots \\ 0 & otherwise. \end{cases} \tag{2.31}$$

For simplicity of the expression we will assume that for the new time indices in the expanded signal, the amplitude values are 0, so we will not always explicitly write the second condition in (2.31), i.e., we will only use $y[n] = x\left[\frac{n}{L}\right]$ to describe the signal expansion.

To draw the graph of $y[n] = x\left[\frac{n}{L}\right]$, or to obtain the expanded signal, $y[n] = x\left[\frac{n}{L}\right]$ we divide the time axis of $x[n]$ by $1/L$, i.e., we multiply the time axis of $x[n]$ by L. This operation is illustrated with an example now.

Example 2.12 If $x[n] = \begin{bmatrix} 1 & 3 & 5 & 7 & 9 & \underset{n=0}{11} & 13 & 15 & 17 \end{bmatrix}$ find $y[n] = x\left[\frac{n}{3}\right]$.

Solution 2.12 The indices for amplitude values of $x[n]$ are explicitly written in

$$x[n] = \begin{bmatrix} \underset{n=-5}{1} & \underset{n=-4}{3} & \underset{n=-3}{5} & \underset{n=-2}{7} & \underset{n=-1}{9} & \underset{n=0}{11} & \underset{n=1}{13} & \underset{n=-2}{15} & \underset{n=3}{17} \end{bmatrix}.$$

Dividing the indices of $x[n]$ by $1/3$, i.e., multiplying the indices by 3, we get the sequence

$$\begin{bmatrix} \underset{n=-15}{1} & \underset{n=-12}{3} & \underset{n=-9}{5} & \underset{n=-6}{7} & \underset{n=-3}{9} & \underset{n=0}{11} & \underset{n=3}{13} & \underset{n=6}{15} & \underset{n=9}{17} \end{bmatrix}.$$

Inserting missing indices and inserting 0 for amplitudes of the missing indices, we obtain the signal $y[n]$ as

$$y[n] = \begin{bmatrix} 1 & 0 & 0 & 3 & 0 & 0 & 5 & 0 & 0 & 7 & 0 & 0 & 9 & 0 & 0 & \underset{n=0}{11} & 0 & 0 & 13 \end{bmatrix}$$

$$\begin{bmatrix} 0 & 0 & 15 & 0 & 0 & 17 \end{bmatrix}.$$

2.2.2 Mathematical Formulization of Upsampling

The upsampling, expansion, of $x[n]$ by L is defined as

$$y[n] = \begin{cases} x\left[\frac{n}{L}\right] & n = 0, \pm L, \pm 2L, \ldots \\ 0 & otherwise \end{cases} \tag{2.32}$$

which can be written in terms of impulse function as

$$y[n] = \sum_{k=-\infty}^{\infty} x[k]\delta[n - kL]. \qquad (2.33)$$

When the summation in (2.33) is expanded, we obtain

$$y[n] = \cdots + x[-1]\delta[n+L] + x[0]\delta[n] + x[1]\delta[n-L] + \cdots$$

Note that to find $x\left[\frac{n}{L}\right]$, we simply insert $L - 1$ zeros between two samples of $x[n]$, that is, if

$$x[n] = [a \quad b \quad c \quad d \quad e],$$

then to get $x\left[\frac{n}{4}\right]$ simply insert 3 zeros between every two samples of $x[n]$, and this operation yields

$$x\left[\frac{n}{4}\right] = [a \quad 0 \quad 0 \quad 0 \quad b \quad 0 \quad 0 \quad 0 \quad c \quad 0 \quad 0 \quad 0 \quad d \quad 0 \quad 0 \quad 0 \quad e].$$

2.2.3   *Frequency Domain Analysis of Upsampling*

Let's try to find the Fourier transform of

$$y[n] = \begin{cases} x\left[\frac{n}{L}\right] & n = 0, \pm L, \pm 2L, \ldots \\ 0 & otherwise. \end{cases} \qquad (2.34)$$

For this purpose, let's start with the definition of the Fourier transform of $y[n]$

$$Y_n(w) = \sum_{n=-\infty}^{\infty} y[n]e^{-jwn} \qquad (2.35)$$

where substituting $\sum_{k=-\infty}^{\infty} x[k]\delta[n - kL]$ for $y[n]$, we get

$$Y_n(w) = \sum_{n=-\infty}^{\infty} \sum_{k=-\infty}^{\infty} x[k]\delta[n - kL]e^{-jwn} \qquad (2.36)$$

in which changing the order of summation terms, we obtain

$$Y_n(w) = \sum_{k=-\infty}^{\infty} \sum_{n=-\infty}^{\infty} x[k]\delta[n - kL]e^{-jwn} \qquad (2.37)$$

which can be rearranged as

$$Y_n(w) = \sum_{k=-\infty}^{\infty} x[k] \underbrace{\sum_{n=-\infty}^{\infty} \delta[n - kL]e^{-jwn}}_{e^{-jwkL}} \tag{2.38}$$

yielding the expression

$$Y_n(w) = \sum_{k=-\infty}^{\infty} x[k]e^{-jwkL}. \tag{2.39}$$

If (2.39) is compared to the Fourier transform of $x[n]$

$$X_n(w) = \sum_{n=-\infty}^{\infty} x[n]e^{-jwn} \tag{2.40}$$

it is seen that

$$Y_n(w) = X_n(Lw) \tag{2.41}$$

Referring to (2.41), it is understood that the graph of $Y_n(w)$ can be obtained by dividing the frequency axis of $X_n(w)$ by L. As it is clear from (2.41) that the spectrum of the upsampled signal gets compressed in frequency domain. In fact, if a signal is expanded in time domain, it is compressed in frequency domain, similarly, if a signal is compressed in time domain, its spectrum expands in frequency domain.

Example 2.13 One period of the Fourier transform of $x[n]$ around origin is given in Fig. 2.52. Draw one period of the Fourier transform of $y[n] = x\left[\frac{n}{L}\right]$.

Solution 2.13 Dividing the frequency axis of $X_n(w)$ by L, we obtain the Fourier transform of $y[n]$ which is depicted in Fig. 2.53.

Note: Don't forget that the Fourier transforms $X_n(w)$ and $Y_n(w)$ are periodic functions with common period 2π. In fact, the Fourier transform of any digital signal is a periodic function with period 2π regardless whether the digital signal is periodic or not in time domain. If the digital signal is periodic in time domain then

Fig. 2.52 One period of the Fourier transform of a digital signal

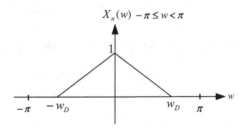

Fig. 2.53 One period of the Fourier transform of upsampled signal for Example 2.12

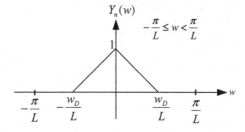

its Fourier transform is an impulse train with period 2π, i.e., its Fourier transform is a discrete signal.

Example 2.14 One period of the Fourier transform of $x[n]$ around origin is given in Fig. 2.54. Draw one period of the Fourier transform of $y[n] = x\left[\frac{n}{2}\right]$.

Solution 2.14 Dividing the frequency axis of $X_n(w)$ by 2, we get the graph in Fig. 2.55 for the Fourier transform of $y[n]$.

To get the graph in Fig. 2.55, we divided the horizontal axis of $X_n(w)$ by 2. Since $Y_n(w)$ is a periodic function with period 2π, the graph in Fig. 2.55 can also be drawn for the interval $-\pi \le w < \pi$ as shown in Fig. 2.56.

Example 2.15 For the system given in Fig. 2.44 $M = L = 2$, and

$$x[n] = [\underset{n=0}{\underbrace{1}} \quad 2 \quad 3 \quad 4 \quad 5 \quad 6 \quad 7 \quad 8 \quad 9 \quad 10].$$

Find the signals $x_d[n]$ and $y[n]$ in Fig. 2.57.

Fig. 2.54 One period of the Fourier transform of a digital signal

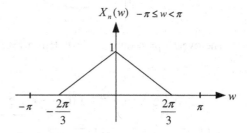

Fig. 2.55 One period of the Fourier transform of upsampled signal for Example 2.13

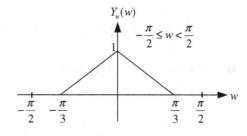

Fig. 2.56 One period of the
Fourier transform of
upsampled signal for Example
2.13

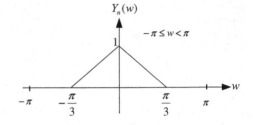

Fig. 2.57 Signal processing
system for Example 2.14

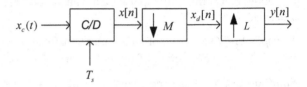

Solution 2.15 To find $x_d[n]$, we divide the time indices of $x[n]$ by 2 and keep only integer division results. This operation yields

$$x_d[n] = [\underbrace{1}_{n=0} \quad 3 \ 5 \ 7 \ 9].$$

To find $y[n]$, we divide the time indices of $x_d[n]$ by $\frac{1}{2}$, i.e., multiply the time indices of $x_d[n]$ by 2. For new indices, amplitude values are equated to 0. The result of this operation is the signal

$$y[n] = [\underbrace{1 \ 0}_{n=0} \quad 3 \ 0 \ 5 \ 0 \ 7 \ 0 \ 9].$$

The overall procedure is illustrated in Fig. 2.58.

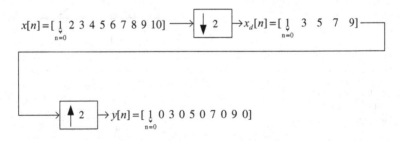

Fig. 2.58 Downsampling and upsampling

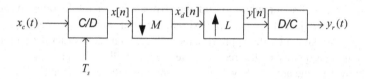

Fig. 2.59 Signal processing system

2.2.4 Interpolation

Let's consider the signal processing system shown in Fig. 2.59. The system includes one downsampler, one upsampler and one D/C converter. Let's study the reconstructed signal $y_r(t)$.

Assume that $y[n]$ is a causal signal. The signal $y_r(t)$ is calculated from the digital signal $y[n]$ using

$$y_r(t) = \sum_{n=-\infty}^{\infty} y[n] h_r(t - nT_s) \tag{2.42}$$

where $h_r(t)$ can either be ideal reconstuction filter, i.e., $h_r(t) = sinc(t/T_s)$ or triangular approximated reconstruction filter, or any other approximated filter. When we expand the summation in (2.42), we see that some of the shifted filters are multiplied by 0, since some of the samples of $y[n]$ are 0. The expansion of (2.42) happens to be as

$$y_r(t) = y[0] h_r(t) + y[1] h_r(t - T_s) + y[2] h_r(t - 2T_s) + y[3] h_r(t - 3T_s) + \cdots \tag{2.43}$$

yielding

$$y_r(t) = 1 \times h_r(t) + 0 \times h_r(t - T_s) + 3 \times h_r(t - 2T_s) + 0 \times h_r(t - 3T_s) + \cdots \tag{2.44}$$

Multiplication of some of the shifted filters by 0 results in information loss in the reconstructed signal.

Question: So how can we increase the quality of the reconstructed signal?

Answer: If we can replace 0 values in the expanded signal $y[n]$ by their estimated values, $y_r(t)$ expression in (2.44) will not include 0 multiplication terms and reconstructed signal becomes better. That is,

$$x[n] = [\underset{n=0}{1}\ 2\ 3\ 4\ 5\ 6\ 7\ 8\ 9\ 10]$$

$$y[n] = [\underset{n=0}{1}\ 0\ 3\ 0\ 5\ 0\ 7\ 0\ 9\ 0]$$

Replace 0's by the estimated values of the omitted samples

Omitted samples are 2, 4, 6, 8, 10

So how can we find a method to find approximate values for the omitted samples of original signal $x[n]$? If we can approximate omitted samples, we can replace 0's in the expanded signal by the approximated values, then reconstruct the continuous time signal. The quality of the reconstructed signal will be better.

We know that the amplitude values of a continuous time signal at time instants t_i and t_{i+1} does not change sharply. Otherwise, it violates the definition of continuous time signal. For instance, the amplitude values of a continuous time signal for three time instants are given in Fig. 2.60.

Hence for the omitted samples, we can make a linear estimation. Assume that $L = M = 2$, in this case, during the downsampling operation; we omit one sample from every other 2 samples. After upsampling operation, we have 0 in the place of omitted sample. We can estimate the omitted sample using the neighbor samples of the omitted sample.

In Fig. 2.60, assume that after sampling operation, we obtain the digital signal $[a\ b\ c]$, and in this case, downsampled signal can be calculated as $[a\ \ c]$. The expanded signal or upsampled signal becomes as $[a\ \ 0\ \ c]$ where 0 can be replaced by the estimated value $\frac{a+c}{2}$. In general if there are $L-1$ zeros between two samples of the expanded signal, we can estimate the omitted samples drawing a line between the amplitudes of these two samples as illustrated in Fig. 2.61.

The missing samples in Fig. 2.61. can be calculated using

$$y[n_i] = b + \frac{a-b}{L}(n_{k+L-1} - n_i), \quad i = k : k+L-1. \tag{2.45}$$

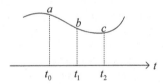

Fig. 2.60 Amplitude values of a continuous time signal for three distinct time instants

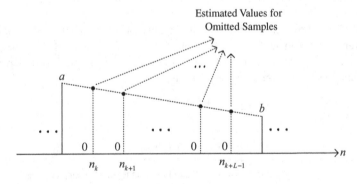

Fig. 2.61 Linear estimation of the missing samples

Let $\Delta = \frac{a-b}{L}$, when (2.45) is expanded for $i = k : k+L-1$, we get the amplitude vector

$$[b+(L-1)\Delta \quad b+(L-2)\Delta \cdots b+2\Delta \quad b+\Delta]. \tag{2.46}$$

Example 2.16 Let $x[n] = [\underset{n=0}{\underbrace{1}} \quad 2 \quad 5 \quad 7 \quad 9 \quad 10 \quad 10]$ find the signals $x_d[n] = x[3n] \ y[n] = x_d \left[\frac{n}{3}\right]$ and using linear estimation method, estimate the missing samples in $y[n]$.

Solution 2.16 To calculate the downsampled signal, we divide the time axis of $x[n]$ by 3 and keep only integer division results, and in a similar manner, to calculate the upsampled signal, we multiply the time axis of the downsampled signal by 3, and for the new time instants 0's are assigned for amplitude values. The downsampled and upsampled signals can be calculated as

$$x_d[n] = [\underset{n=0}{\underbrace{1}} \quad 7 \quad 10] \quad y[n] = [\underset{n=0}{\underbrace{1}} \quad 0 \quad 0 \quad 7 \quad 0 \quad 0 \quad 10].$$

and these signals are graphically shown in Fig. 2.62.
The missing samples in upsampled signal can be calculated using

$$\Delta = \frac{a-b}{L}, \quad \text{and} \quad [b+(L-1)\Delta \quad b+(L-2)\Delta \quad \cdots \quad b+2\Delta \quad b+\Delta]$$

For the first 2 missing samples

$$\Delta = \frac{1-7}{3} \rightarrow \Delta = -2$$

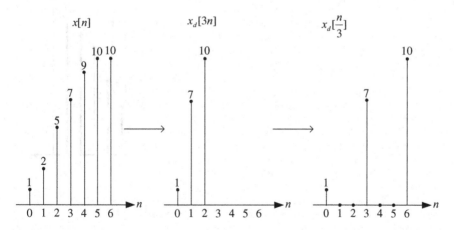

Fig. 2.62 Original signal, downsampled signal, upsampled signal

and the missing samples are

$$[7+2(-2) \quad 7+1(-2)] \rightarrow [3 \quad 4].$$

For the next 2 missing samples

$$\Delta = \frac{7-10}{3} \rightarrow \Delta = -1$$

and the missing samples are

$$[10+2(-1) \quad 10+1(-1)] \rightarrow [8 \quad 9].$$

The calculation of the missing samples is graphically illustrated in Fig. 2.63. Hence with the estimated values, the upsampled signal becomes as

$$y[n] = [\underbrace{1}_{n=0} \quad 3 \quad 4 \quad 7 \quad 8 \quad 9 \quad 10]. \tag{2.47}$$

The original sequence before downsampling operation was

$$x[n] = [\underbrace{1}_{n=0} \quad 2 \quad 5 \quad 7 \quad 9 \quad 10 \quad 10]. \tag{2.48}$$

When (2.47) is compared to (2.48), we see that the calculated samples are close to the original omitted samples.

Fig. 2.63 Calculation of the missing samples

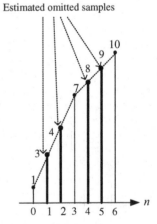

Estimated omitted samples

2.2.5   Mathematical Analysis of Interpolation

We explained an estimation method for the calculation of missing samples in expanded signal. However, we did not follow a mathematical analysis. How can we find the missing samples in upsampled (expanded) signal using a mathematical approach?

In time domain, it is difficult to find a mathematical approach for the estimation of missing samples. Let's approach to the problem in frequency domain. Let's consider the system involving downsampling and upsampling operations given in Fig. 2.64 where we assume that $L = M$.

Let's assume that the Fourier transform of $x[n]$ is as in Fig. 2.65. We will inspect the Fourier transforms of $y[n]$ and $x[n]$ in Fig. 2.64 and find a relation between them.

Considering Fig. 2.65 the Fourier transform of $x_d[Mn]$ can be drawn as in Fig. 2.66.

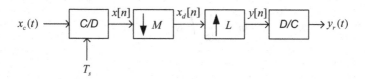

Fig. 2.64 Signal processing system including upsampling and downsampling operations

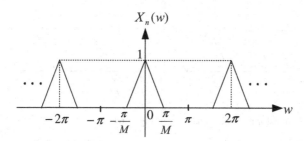

Fig. 2.65 Fourier transform of a digital signal

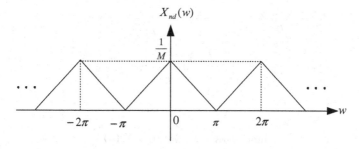

Fig. 2.66 Fourier transform of $x_d[Mn]$

Dividing the horizontal axis of the graph in Fig. 2.66 by L, we obtain the graph of $Y_n(w)$ as Fig. 2.67.

If we compare the graph of $X_n(w)$ in Fig. 2.65 to the graph of $Y_n(w)$ in Fig. 2.67, it is seen that for $\frac{\pi}{L} \leq |w| < 2\pi - \frac{\pi}{L}$ $X_n(w) = 0$ but $Y_n(w) \neq 0$, and for other frequency intervals, $Y_n(w) = \frac{1}{M}X_n(w)$. This is illustrated in Fig. 2.68.

How can we make $Y_n(w)$ to be equal to $X_n(w)$ for all frequency values? This is possible if we multiply $Y_n(w)$ by a lowpass digital filter with the transfer function as in Fig. 2.69.

Since $L = M$ and $Y_i(w) = H_i(w)Y_n(w)$, we can show the multiplication of $H_i(w)Y_n(w)$ as in Fig. 2.70.

The result of the above multiplication is depicted in Fig. 2.71.

For $L = M$, we have $Y_i(w) = X_n(w)$ which means that $y_i[n] = x[n]$, that is omitted samples are reconstructed perfectly.

Let's now do the time domain analysis of this reconstruction process. If $Y_i(w) = H_i(w)Y_n(w)$, then $y_i[n] = h_i[n] * y[n]$. The time domain expression $h_i[n]$ can be obtained via inverse Fourier transform

$$h_i[n] = \frac{1}{2\pi} \int_{2\pi} H_i(w)e^{jwn} dw \qquad (2.49)$$

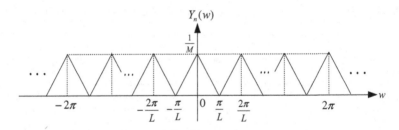

Fig. 2.67 Fourier transform of the signal $y[n]$ in Fig. 2.64

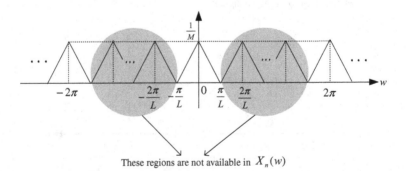

These regions are not available in $X_n(w)$

Fig. 2.68 Comparison of $X_n(w)$ and $Y_n(w)$

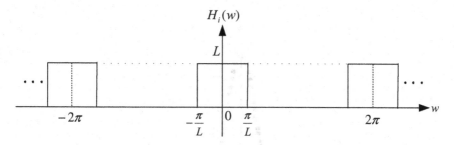

Fig. 2.69 Lowpass digital filter

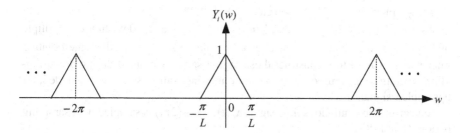

Fig. 2.70 The multiplication of $H_i(w)Y_n(w)$

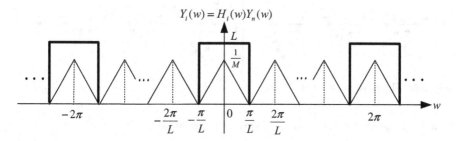

Fig. 2.71 The graph of $Y_i(w) = H_i(w)Y_n(w)$

where using the frontiers $-\frac{\pi}{L}, \frac{\pi}{L}$, we get

$$h_i[n] = \frac{1}{2\pi} \int_{-\frac{\pi}{L}}^{\frac{\pi}{L}} L e^{jwn} dw \rightarrow h_i[n] = \frac{\sin\left(\frac{\pi n}{L}\right)}{\frac{\pi n}{L}} \tag{2.50}$$

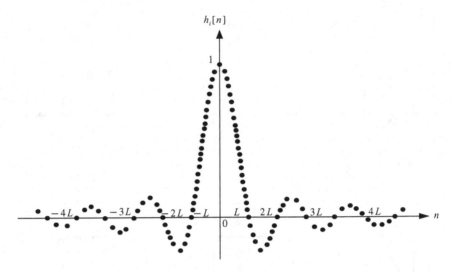

Fig. 2.72 The graph of $\sin c(n/L)$

which can be expressed in terms of $\sin c(\cdot)$ function as

$$h_i[n] = \sin c\left(\frac{n}{L}\right). \tag{2.51}$$

The graph of $\sin c(n/L)$ is depicted in Fig. 2.72.

As it is seen from Fig. 2.72 that $h_i[n] = \sin c\left(\frac{n}{L}\right)$ equals to 0 when n is a multiple of L. The digital filter with impulse response $h_i[n] = \sin c\left(\frac{n}{L}\right)$ is called interpolating filter which is used to reconstruct those digital samples omitted during downsampling operation, i.e., used to reconstruct missing samples in the expanded, or upsampled signal.

Exercise: The continuous time signal $x_c(t) = \cos(2\pi t)$ is sampled with sampling period $T_s = 1/8$ s.

(a) For a mathematical sequence $x[n]$ from the samples taken from continuous time signal in the interval 0–1 s.
(b) $x[n]$ is downsampled by $M = 2$, and $x_d[n]$ is the downsampled signal, find $x_d[n]$.
(c) The downsampled signal $x_d[n]$ is upsampled and let $y[n]$ be the upsampled signal, find $y[n]$.
(d) Calculate the missing samples in $y[n]$ using the ideal interpolation filter.

2.2.6 Approximation of the Ideal Interpolation Filter

Since digital $\sin c(\cdot)$ filter is an ideal filter, it is difficult to implement such filters, instead we can use an approximation of this digital filter. As it is clear from Fig. 2.72, the digital $\sin c(\cdot)$ filter includes a large main lobe centered upon origin, and many other side lobes. To approximate the digital $\sin c(\cdot)$ filter, we can use triangles for the lobes in Fig. 2.72. The simplest approximation is to use an isosceles triangle for the main lobe and omit the other side lobes.

The simplest approximated digital can filter can be obtained as shown in Fig. 2.73.

Referring to Fig. 2.73 the approximated interpolation filter can mathematically be expressed as

$$h_{ai}[n] = \begin{cases} \frac{n}{L} + 1, & if \quad -L \leq n < 0 \\ -\frac{n}{L} + 1, & if \quad 0 \leq n < L \\ 0, & otherwise \end{cases} \tag{2.52}$$

which can be expressed in more compact form as

$$h_{ai}[n] = \begin{cases} -\frac{|n|}{L} + 1, & if \quad -L \leq n < L \\ 0, & otherwise. \end{cases} \tag{2.53}$$

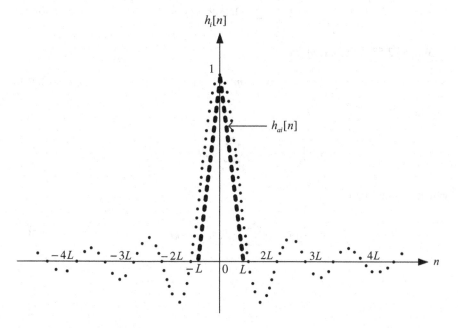

Fig. 2.73 Approximation of the ideal interpolation filter

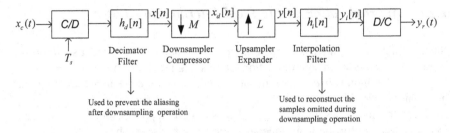

Fig. 2.74 Signal processing system with interpolation filter

With the interpolation filter our complete signal processing system becomes as in Fig. 2.74.

For the reconstruction of the samples omitted during downsampling operation, if approximated interpolating filter is used, the reconstructed digital signal can be written as

$$y_i[n] = h_{ai}[n] * y[n] \rightarrow y_i[n] = \sum_{k=-\infty}^{\infty} y[k]h_{ai}[n-k] \qquad (2.54)$$

where $h_{ai}[n]$ denotes the approximated reconstruction filter, or interpolation filter. Now let's try to write a relation between $x_d[n]$ and $y_i[n]$ given in Fig. 2.74. We know that

$$y[n] = \sum_{k=-\infty}^{\infty} x_d[k]\delta[n-kL]. \qquad (2.55)$$

When (2.53) is replaced into

$$y_i[n] = h_i[n] * y[n] \qquad (2.56)$$

we get

$$y_i[n] = h_i[n] * \sum_{k=-\infty}^{\infty} x_d[k]\delta[n-kL] \qquad (2.57)$$

which is simplified as

$$y_i[n] = \sum_{k=-\infty}^{\infty} x_d[k]h_i[n-kL]. \qquad (2.58)$$

When (2.58) is expanded, we get the explicit form of $y_i[n]$ as

$$y_i[n] = \cdots + x_d[-1]h_i[n+L] + x_d[0]h_i[n] + x_d[1]h_i[n-L] + \cdots \quad (2.59)$$

Using the ideal interpolation filter, i.e., ideal reconstruction filter,

$$h_i[n] = \frac{\sin\left(\frac{\pi n}{L}\right)}{\frac{\pi n}{L}}$$

in (2.58), we can write the reconstructed digital signal as

$$y_i[n] = \sum_{k=-\infty}^{\infty} x_d[k] \frac{\sin\left(\frac{\pi(n-kL)}{L}\right)}{\frac{\pi(n-kL)}{L}} \quad (2.60)$$

or in terms of $\sin c(\cdot)$ function, we can write (2.60) as

$$y_i[n] = \sum_{k=-\infty}^{\infty} x_d[k] \sin c\left(\frac{n-kL}{L}\right). \quad (2.61)$$

Note: Digital reconstructed signal expression $y_i[n] = \sum_{k=-\infty}^{\infty} x_d[k]h_i[n-kL]$ is quite similar to the analog reconstructed signal expression $x_r(t) = \sum_{k=-\infty}^{\infty} x[k] h_r(t-kT_s)$.

Example 2.17 For the system given in Fig. 2.75 $L = M = 3$ and $x[n] = [1 \quad 2 \quad 3 \quad 4]$. Find $x_d[n], y[n]$, and $y_i[n]$. Use approximated linear digital filter for $h_i[n]$.

Solution 2.17 For $L = M = 3$, if $x[n] = [1 \quad 2 \quad 3 \quad 4]$, then $x_d[n] = [1 \quad 4]$ and $y[n] = [1 \quad 0 \quad 0 \quad 4]$.

To find $y_i[n]$ we can use either

$$y_i[n] = \sum_{k=-\infty}^{\infty} y[k]h_{ai}[n-k] \quad (2.62)$$

or

$$y_i[n] = \sum_{k=-\infty}^{\infty} x_d[k]h_i[n-kL] \quad (2.63)$$

Let's use both of them separately. First using (2.53), let's calculate and draw the linear approximated digital interpolation filter as in Fig. 2.76.

Fig. 2.75 Signal processing system for Example 2.16

Fig. 2.76 Approximated
interpolation filter

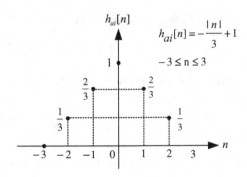

Expanding (2.62), we get

$$y_i[n] = y[0]h_{ai}[n] + y[1]h_{ai}[n-1] + y[2]h_{ai}[n-2] + y[3]h_{ai}[n-3]. \quad (2.64)$$

If $y[n] = \begin{bmatrix} 1 & 0 & 0 & 4 \end{bmatrix}$ is considered, we see that the amplitude values at indices $n = 1$, and $n = 2$, are missing. When $n = 1$ is placed into (2.64), we get

$$y_i[1] = \underbrace{y[0]}_{1}\underbrace{h_{ai}[1]}_{2/3} + \underbrace{y[1]}_{0}\underbrace{h_{ai}[0]}_{1} + \underbrace{y[2]}_{0}\underbrace{h_{ai}[-1]}_{2/3} + \underbrace{y[3]}_{4}\underbrace{h_{ai}[-2]}_{1/3} \quad (2.65)$$

which yields

$$y_i[1] = \frac{2}{3} + \frac{4}{3} \rightarrow y_i[1] = 2 \quad (2.66)$$

and when $n = 2$ is placed into (2.64), we obtain

$$y_i[2] = \underbrace{y[0]}_{1}\underbrace{h_{ai}[2]}_{1/3} + \underbrace{y[1]}_{0}\underbrace{h_{ai}[1]}_{1} + \underbrace{y[2]}_{0}\underbrace{h_{ai}[0]}_{2/3} + \underbrace{y[3]}_{4}\underbrace{h_{ai}[-1]}_{2/3} \quad (2.67)$$

which yields

$$y_i[2] = \frac{1}{3} + \frac{8}{3} \rightarrow y_i[2] = 3 \quad (2.68)$$

So missing samples are found as $y_i[1] = 2$ and $y_i[2] = 3$, and when these samples are replaced by 0's in $y[n]$, we get

$$y_i[n] = \begin{bmatrix} 1 & 2 & 3 & 4 \end{bmatrix}$$

Now let's use the formula

$$y_i[n] = \sum_{k=-\infty}^{\infty} x_d[k]h_i[n-kL].$$ (2.69)

When (2.69) is expanded, noting that $x_d[n] = \begin{bmatrix} 1 & 4 \end{bmatrix}$ and $L = 3$, we get

$$y_i[n] = x_d[0]h_{ai}[n] + x_d[1]h_{ai}[n-3].$$ (2.70)

When (2.70) is evaluated for $n = 1$, we obtain

$$y_i[1] = \underbrace{x_d[0]}_{1}\underbrace{h_{ai}[1]}_{2/3} + \underbrace{x_d[1]}_{4}\underbrace{h_{ai}[-2]}_{1/3}$$

which yields

$$y_i[1] = \frac{2}{3} + \frac{4}{3} \rightarrow y_i[1] = 2$$ (2.71)

and when (2.69) is evaluated for $n = 2$, we get

$$y_i[2] = \underbrace{x_d[0]}_{1}\underbrace{h_{ai}[2]}_{1/3} + \underbrace{x_d[1]}_{4}\underbrace{h_{ai}[-1]}_{2/3}$$ (2.72)

which yields

$$y_i[2] = \frac{1}{3} + \frac{8}{3} \rightarrow y_i[2] = 3.$$ (2.73)

Hence, both formulas give the same results. In addition, we had already introduced the linear estimation method using the continuity property of analog signals. It is now very clear that the linear estimation method is nothing but the use of triangle approximated digital reconstruction filter.

Example 2.18 Show that the systems given in Fig. 2.77 have the same outputs for the same inputs.

Fig. 2.77 Signal processing systems for Example 2.17

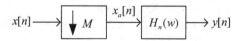

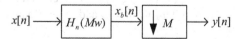

Solution 2.18 For the first system we have

$$X_{an}(w) = \frac{1}{M} \sum_{k=0}^{M-1} X_n \left(\frac{w - k2\pi}{M} \right) \tag{2.74}$$

and

$$Y_n(w) = H_n(w)X_{an}(w) \rightarrow Y_n(w) = \frac{H_n(w)}{M} \sum_{k=0}^{M-1} X_n \left(\frac{w - k2\pi}{M} \right) \tag{2.75}$$

For the second system we have

$$X_{bn}(w) = H_n(Mw)X_n(w) \tag{2.76}$$

and

$$Y_n(w) = \frac{1}{M} \sum_{k=0}^{M-1} X_{bn} \left(\frac{w - k2\pi}{M} \right). \tag{2.77}$$

When (2.76) is inserted into (2.77), we obtain

$$Y_n(w) = \frac{1}{M} \sum_{k=0}^{M-1} H_n \left(M \frac{w - k2\pi}{M} \right) X_n \left(\frac{w - k2\pi}{M} \right). \tag{2.78}$$

Since $H_n(w)$ is a periodic function with period 2π, (2.78) can be written as

$$Y_n(w) = \frac{1}{M} \sum_{k=0}^{M-1} H_n(w)X_n \left(\frac{w - k2\pi}{M} \right) \tag{2.79}$$

which is equal to

$$Y_n(w) = H_n(w) \frac{1}{M} \sum_{k=0}^{M-1} X_n \left(\frac{w - k2\pi}{M} \right) \rightarrow Y_n(w) = H_n(w)X_{an}(w). \tag{2.80}$$

When (2.75) is compared to (2.80), we see that both systems have the same outputs for the same inputs.

Exercise: Show that the systems given below have the same outputs for the same inputs (Fig. 2.78).

Example 2.19 For the system given in Fig. 2.79, find a relation in time domain between system input $x[n]$ and system output $y[n]$.

Fig. 2.78 Signal processing
system for exercise

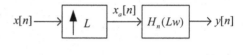

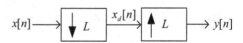

Fig. 2.79 Signal processing
system for Example 2.18

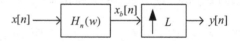

Solution 2.19 We have $x_d[n] = x[Ln]$ and $y[n] = x_d\left[\frac{n}{L}\right]$. Putting $x_d[n]$ expression into $y[n]$ expression, we get $y[n] = x\left[\frac{Ln}{L}\right] \rightarrow y[n] = x[n]$. However, this is not always correct. Since we know that for $L = 2$ if $x[n] = \begin{bmatrix} 1 & 2 & 3 \end{bmatrix}$, then $x_d[n] = \begin{bmatrix} 1 & 3 \end{bmatrix}$ and $y[n] = \begin{bmatrix} 1 & 0 & 3 \end{bmatrix}$, it is obvious that $x[n] \neq y[n]$.

But using $x_d[n] = x[Ln]$ and $y[n] = x_d\left[\frac{n}{L}\right]$, we found $y[n] = x[n]$. So, what is wrong with our approach to the problem?

Because, we did not pay attention to the criteria in upsampling operation. That is, $y[n] = x_d\left[\frac{n}{L}\right]$ if $n = kL, k \in Z$; otherwise, $y[n] = 0$. Then $y[n] = x[n]$ is valid only for some values of n and these n values are multiples of L. That is for $L = 2$ if $x[n] = \begin{bmatrix} 1 & 2 & 3 \end{bmatrix}$, then $x_d[n] = \begin{bmatrix} 1 & 3 \end{bmatrix}$ and $y[n] = \begin{bmatrix} 1 & 0 & 3 \end{bmatrix}$, and $y[n] = x[n]$ for $n = 0, 2$ only.

However, for some signals, no information loss occurs after compression operation. This is possible if the omitted samples are also zeros. In this case, expanded signal equals to the original signal. For example, if

$$x[n] = [\underset{n=0}{a} \quad 0 \quad b \quad 0 \quad c \quad 0 \quad d]$$

then after downsampling by $L = 2$, we get

$$x_d[n] = \begin{bmatrix} a & b & c & d \end{bmatrix}$$

and after expansion by $L = 2$, we obtain

$$y[n] = [\underset{n=0}{a} \quad 0 \quad b \quad 0 \quad c \quad 0 \quad d]$$

Thus, we see that $y[n] = x[n]$ for every n values.

To write a mathematical expression between $x[n]$ and $y[n]$, let's express $x_d[n]$ in terms of $x[n]$ as

$$x_d[n] = \sum_{n=-\infty}^{\infty} x[n] \sum_{r=-\infty}^{\infty} \delta[n - rM] \tag{2.81}$$

and express $y[n]$ in terms of $x_d[n]$ as

$$y[n] = \sum_{k=-\infty}^{\infty} x_d[k]\delta[n - kL]. \tag{2.82}$$

Inserting (2.81) into (2.82), we obtain

$$y[n] = \sum_{k=-\infty}^{\infty} x[k] \sum_{r=-\infty}^{\infty} \delta[k - rM] \sum_{n=-\infty}^{\infty} \delta[n - kL] \tag{2.83}$$

which is the final expression showing the relation between $x[n]$ and $y[n]$.

Example 2.20 Find a method to check whether information loss occurs or not after downsampling by M.

Solution 2.20 If $x[n]$ is downsampled by M, we omit $M - 1$ samples from every M samples. If we denote the information bit indices by the numbers $0, 1, 2, \ldots, M \ldots$, then the first omitted samples have indices $1, 2, \ldots, M - 1$ and the second set of omitted indices have indices $M + 1, M + 2, \ldots, 2M - 1$, and so on.

Hence, by summing the absolute values of the omitted samples and checking whether it equals to zero or not, we can conclude whether information loss occurs or not after downsampling operation. That is, we calculate

$$Loss = \sum_{k=-\infty}^{\infty} \sum_{n=1}^{M-1} |x[n + kM]| \tag{2.84}$$

and if $Loss \neq 0$, then information loss occurs after downsampling of $x[n]$, otherwise not.

Example 2.21 If

$$y[n] = \begin{cases} x[n] & \text{if } n \text{ is even} \\ 0 & \text{otherwise} \end{cases} \tag{2.85}$$

then write a mathematical expression between $x[n]$ and $y[n]$.

Solution 2.21 Using (2.85), we can express $y[n]$ in terms of $x[n]$ as

$$y[n] = \frac{1 + (-1)^n}{2} x[n]. \tag{2.86}$$

Fig. 2.80 Signal processing system for Example 2.21

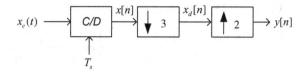

Since $\cos(\pi n) = (-1)^n$, then (2.86) can also be written as

$$y[n] = \frac{1 + \cos(\pi n)}{2} x[n].$$

Example 2.22 For the system given in Fig. 2.80, $x_c(t) = \cos(2000\pi t)$, $T_s = \frac{1}{4000}$ sec find $x[n], x_d[n]$ and $y[n]$.

Solution 2.22 When continuous time signal is sampled, we get

$$x[n] = x_c(t)|_{t=nT_s} \rightarrow x[n] = \cos\left(2000\pi n \frac{1}{4000}\right) \rightarrow x[n] = \cos\left(\frac{\pi}{2}n\right). \quad (2.87)$$

After downsampling operation, we have

$$x_d[n] = x[3n] \rightarrow x_d[n] = \cos\left(\frac{3\pi}{2}n\right) \quad (2.88)$$

After upsampling operation, we have

$$y[n] = \begin{cases} x_d\left[\frac{n}{2}\right] & n \text{ is even} \\ 0 & \text{otherwise} \end{cases} \quad (2.89)$$

which yields

$$y[n] = \begin{cases} \cos\left(\frac{\pi}{4}n\right) & n \text{ is even} \\ 0 & \text{otherwise} \end{cases} \quad (2.90)$$

The mathematical expression in (2.90) can be written in a more compact manner as

$$y[n] = \frac{1 + \cos(\pi n)}{2} \cos\left(\frac{\pi}{4}n\right). \quad (2.91)$$

Using the property

$$\cos(a)\cos(b) = \frac{1}{2}(\cos(a+b) + \cos(a-b)) \quad (2.92)$$

Equation (2.91) can be written as

$$y[n] = \frac{1}{2}\cos\left(\frac{\pi}{4}n\right) + \frac{1}{4}\cos\left(\frac{5\pi}{4}n\right) + \frac{1}{4}\cos\left(\frac{3\pi}{4}n\right) \qquad (2.93)$$

where using $\cos(\theta) = \cos(2\pi - \theta)$ Eq. (2.93) can be written as

$$y[n] = \frac{1}{2}\cos\left(\frac{\pi}{4}n\right) + \frac{1}{2}\cos\left(\frac{3\pi}{4}n\right). \qquad (2.94)$$

Note: $\cos\left(\frac{5\pi}{4}n\right) = \cos\left(2\pi n - \frac{5\pi}{4}n\right) \rightarrow \cos\left(\frac{5\pi}{4}n\right) = \cos\left(\frac{3\pi}{4}n\right)$

Example 2.23 $x_c(t) = e^{jw_N t}$ and $x[n] = x_c(t)|_{t=nT_s}$, $T_s = 1$ find the Fourier transforms of $x_c(t)$ and $x[n]$.

Solution 2.23 The Fourier transform of the continuous time exponential signal is

$$X_c(w) = 2\pi\delta(w - w_N) \qquad (2.95)$$

which is depicted in Fig. 2.81.

If $x[n] = x_c(t)|_{t=nT_s}$, then one period of the Fourier transform of $x[n]$ is

$$X_n(w) = \frac{1}{T_s}X_c\left(\frac{w}{T_s}\right), \qquad |w| < \pi \qquad (2.96)$$

which is shown in Fig. 2.82.

Figure 2.82 can mathematically be expressed as $X_n(w) = 2\pi\delta(w - w_D)$, $|w| < 2\pi$. Since $X_n(w)$ is the Fourier transform of a digital signal, it is a periodic function and its period equals to 2π and it can be written as

$$X_n(w) = 2\pi \sum_{k=-\infty}^{\infty} \delta(w - w_D - k2\pi). \qquad (2.97)$$

Fig. 2.81 Fourier transform of continuous time exponential signal

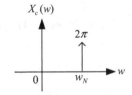

Fig. 2.82 One period of the Fourier transform of digital exponential signal

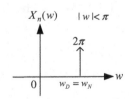

After sampling of the continuous time exponential signal, we obtain

$$x[n] = e^{\overbrace{jw_N T_s n}^{w_D}} \rightarrow x[n] = e^{jw_D n}.$$

Hence we can write the following transform pair in general

$$e^{jw_0 n} \overset{FT}{\leftrightarrow} 2\pi \sum_{k=-\infty}^{\infty} \delta(w - w_0 - k2\pi). \tag{2.98}$$

Example 2.24 Given $x[n] = e^{j\frac{\pi}{3}n}$, find Fourier transform of $x[n]$, i.e., $X_n(w)$.

Solution 2.24 $X_n(w) = 2\pi\delta(w - \frac{\pi}{3})$, $|w| < \pi$ and $X_n(w)$ is periodic with period 2π, so in more compact form, we can write it as

$$X_n(w) = 2\pi \sum_{k=-\infty}^{\infty} \delta(w - \frac{\pi}{3} - k2\pi) \tag{2.99}$$

Example 2.25 $x[n] = \cos(w_0 n)$, $y[n] = \cos(\frac{\pi}{3}n)$, $w[n] = \cos(\frac{2\pi}{3}n)$, find the Fourier transforms of $x[n], y[n]$, and $w[n]$.

Solution 2.25 We know that $\cos(\theta) = \frac{1}{2}\left(e^{j\theta} + e^{-j\theta}\right)$ and $\sin(\theta) = \frac{1}{2j}\left(e^{j\theta} - e^{-j\theta}\right)$, and using the Fourier transform of digital exponential function, we obtain the results

$$X_n(w) = \pi(\delta(w - w_0) + \delta(w + w_0)), \quad |w| < \pi$$

$$Y_n(w) = \pi\left(\delta\left(w - \frac{\pi}{3}\right) + \delta\left(w + \frac{\pi}{3}\right)\right), \quad |w| < \pi$$

$$W_n(w) = \pi\left(\delta\left(w - \frac{2\pi}{3}\right) + \delta\left(w + \frac{2\pi}{3}\right)\right), \quad |w| < \pi.$$

$X_n(w)$, $Y_n(w)$, and $W_n(w)$ are periodic functions with period 2π.

Example 2.26 The transfer function of a lowpass digital filter is depicted in Fig. 2.84. Accordingly, find the output of the block diagram shown in Fig. 2.83 for the input signal

$$x[n] = \cos\left(\frac{\pi}{3}n\right) + \cos\left(\frac{2\pi}{3}n\right).$$

The Fourier transform of the filter impulse is given as in Fig. 2.84.

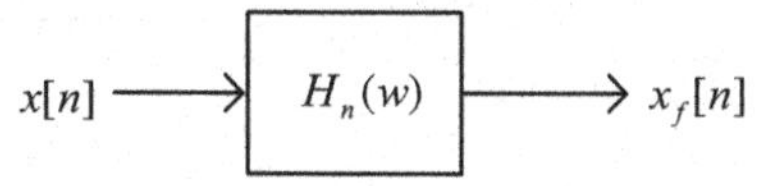

Fig. 2.83 Lowpass filtering of digital signals

$$x[n] \longrightarrow \boxed{H_n(w)} \longrightarrow x_f[n]$$

Fig. 2.84 Digital lowpass filter transfer function

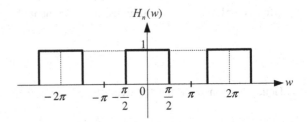

Solution 2.26 If digital frequency w is between $-\frac{\pi}{2}$ and $\frac{\pi}{2}$, that is if $|w| < \frac{\pi}{2}$, the digital frequency is accepted as low frequency. On the other hand, if $\frac{\pi}{2} < |w| < \pi$, the digital frequency is accepted as high frequency.

One period of Fourier transform of $x[n]$ can be calculated as

$$X_n(w) = \pi\left(\delta\left(w - \frac{\pi}{3}\right) + \delta\left(w + \frac{\pi}{3}\right)\right) + \pi\left(\delta\left(w - \frac{2\pi}{3}\right) + \delta\left(w + \frac{2\pi}{3}\right)\right), \quad |w| < \pi$$

$$(2.100)$$

which is graphically illustrated in Fig. 2.85.

At the output of the block diagram, we have $X_{fn}(w) = H_n(w)X_n(w)$ and this multiplication is graphically illustrated in Fig. 2.86.

As it is obvious from Fig. 2.86, the signal $X_{fn}(w) = H_n(w)X_n(w)$ equals to

$$X_{fn}(w) = \pi\left(\delta\left(w - \frac{\pi}{3}\right) + \delta\left(w + \frac{\pi}{3}\right)\right).$$

$$(2.101)$$

Fig. 2.85 Fourier transform of the input signal in Example 2.25

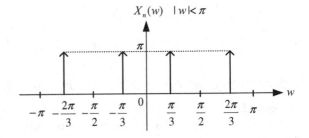

Fig. 2.86 Multiplication of $X_n(w)$ and $H_n(w)$

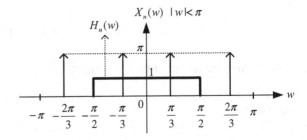

That is, high frequency part of the signal is filtered by the low pass filter, and at the output of the filter, only low frequency components exist. In time domain, the filter output equals to

$$x_f[n] = \cos\left(\frac{\pi}{3}n\right). \tag{2.102}$$

Example 2.27 In the system of Fig. 2.87, $x_c(t) = \cos(2000\pi t) + \cos(5000\pi t)$, $T_s = \frac{1}{3000}$ and transfer function of the digital filter is depicted in Fig. 2.88.

Find $x[n], x_f[n]$, and $x_d[n]$.

Solution 2.27 $x[n] = x_c(t)|_{t=nT_s}$ leads to

$$x[n] = \cos\left(\frac{2\pi}{3}n\right) + \cos\left(\frac{5\pi}{3}n\right). \tag{2.103}$$

Since $\cos\left(\frac{5\pi}{3}n\right) = \cos\left(2\pi n - \frac{5\pi}{3}n\right) \rightarrow \cos\left(\frac{5\pi}{3}n\right) = \cos\left(\frac{\pi}{3}n\right)$, then (2.103) becomes as

$$x[n] = \cos\left(\frac{2\pi}{3}n\right) + \cos\left(\frac{\pi}{3}n\right). \tag{2.104}$$

The digital filter eliminates high frequency component of $x[n]$, hence at the output of the filter we have

$$x_f[n] = \cos\left(\frac{\pi}{3}n\right). \tag{2.105}$$

Fig. 2.87 Signal processing system for Example 2.26

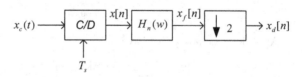

Fig. 2.88 Digital lowpass filter transfer function

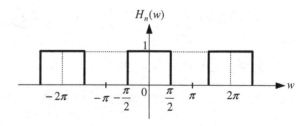

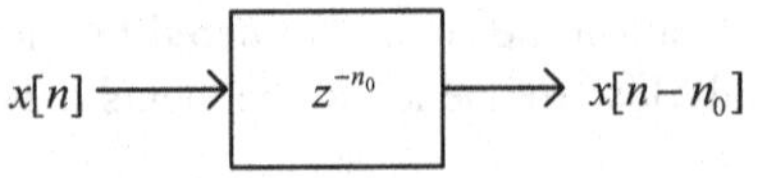

$$x[n] \longrightarrow \boxed{z^{-n_0}} \longrightarrow x[n-n_0]$$

Fig. 2.89 Delay system

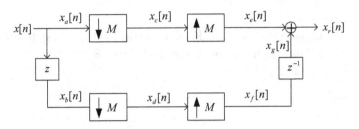

Fig. 2.90 Signal processing system for Example 2.27

After downsampling operation, we get

$$x_d[n] = x_f[2n] \rightarrow x_d[n] = \cos\left(\frac{2\pi}{3}n\right). \tag{2.106}$$

Example 2.28 The delay system is shown in Fig. 2.89.

In the system shown in Fig. 2.90, $M = 2$ and $x[n] = [1 \quad 2 \quad 3 \quad 4 \quad 5 \quad 6]$. Find $x_a[n], x_b[n], x_c[n], x_d[n], x_e[n], x_f[n]$ and $x_r[n]$.

Solution 2.28 If $x[n] = [1 \quad 2 \quad 3 \quad 4 \quad 5 \quad 6]$, then $x_a[n] = [\underset{n=0}{\underbrace{1}} \quad 2 \quad 3 \quad 4$
$5 \quad 6]$ and since $x_b[n] = x[n+1]$ moving $n = 0$ pointer to the right by one unit, we get

$$x_b[n] = [1 \quad \underset{n=0}{\underbrace{2}} \quad 3 \quad 4 \quad 5 \quad 6]$$

After downsampling, we have

$$x_c[n] = [\underset{n=0}{\underbrace{1}} \quad 3 \quad 5] \quad x_d[n] = [\underset{n=0}{\underbrace{2}} \quad 4 \quad 6].$$

After upsampling, we have

$$x_e[n] = [\underset{n=0}{\underbrace{1}} \quad 0 \quad 3 \quad 0 \quad 5] \quad x_f[n] = [\underset{n=0}{\underbrace{2}} \quad 0 \quad 4 \quad 0 \quad 6].$$

After delay operator z^{-1}, we have

$$x_g[n] = [\underset{n=0}{\underbrace{0}} \quad 2 \quad 0 \quad 4 \quad 0 \quad 6].$$

And at the system output, we have

$$x_r[n] = x_e[n] + x_g[n]$$

where

$$x_e[n] = [\underset{n=0}{1} \quad 0 \quad 3 \quad 0 \quad 5] \quad x_g[n] = [\underset{n=0}{0} \quad 2 \quad 0 \quad 4 \quad 0 \quad 6].$$

Hence,

$$x_r[n] = [1 \quad 2 \quad 3 \quad 4 \quad 5 \quad 6].$$

The signal flow of the system in Fig. 2.90 is shown in Fig. 2.91.

Exercise: For the system given in Fig. 2.92, $M = 3$ and

$$x[n] = [1 \quad 2 \quad 3 \quad 4 \quad 5 \quad 6 \quad 7 \quad 8 \quad 9 \quad 10 \quad 11 \quad 12 \quad 13 \quad 14 \quad 15].$$

Find the output of every block and finally find $x_r[n]$.

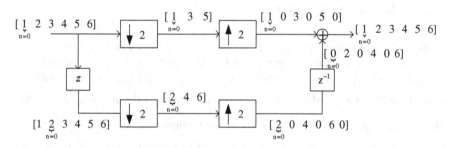

Fig. 2.91 Signal flow for the system in Fig. 2.90

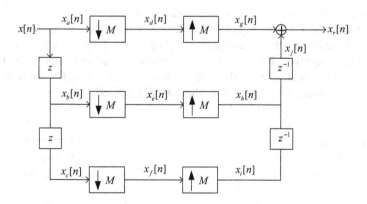

Fig. 2.92 Signal processing system for exercise

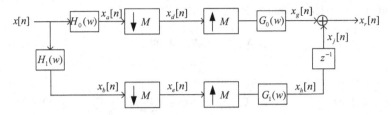

Fig. 2.93 Signal processing system for Example 2.28

Example 2.29 For the system shown in Fig. 2.93, $x[n] = \cos\left(\frac{2\pi}{3}n\right) + \cos\left(\frac{\pi}{3}n\right)$, $M = 2$.

Find $H_0(w), H_1(w), G_0(w)$, and $G_1(w)$ such that $x_r[n] = x[n]$.

Solution 2.29 $H_0(w)$ can be chosen as a low pass digital filter. $H_1(w)$ can be chosen as a high pass digital filter. $G_0(w)$ and $G_1(w)$ are interpolating $\operatorname{sin}c(\cdot)$ filters.

2.2.7 Anti-aliasing Filter

Consider the continuous to digital conversion system shown in Fig. 2.94.

We know that to obtain one period the Fourier transform of $x[n]$, we multiply the frequency axis of the Fourier transform of $x_c(t)$ by T_s and multiply the amplitude axis of the Fourier transform of $x_c(t)$ by $1/T_s$, i.e., we calculate $\frac{1}{T_s}X_c\left(\frac{w}{T_s}\right)$. If the Fourier transform of $x_c(t)$ has a bandwidth greater than π/T_s, then $\frac{1}{T_s}X_c\left(\frac{w}{T_s}\right)$ extends beyond $(-\pi, \pi)$ and aliasing observed in the Fourier transform of $x[n]$. This situation is described in Fig. 2.95.

Since $X_n(w)$ is periodic with period 2π when $\frac{1}{T_s}X_c\left(\frac{w}{T_s}\right)$ extends beyond $(-\pi, \pi)$, overlapping will be observed in $X_n(w)$ as shown in Fig. 2.96.

The portion of $X_n(w)$ in Fig. 2.96 for $|w| < \pi$ is shown in Fig. 2.96.

To decrease the effect of aliasing (overlapping) in the digital signal, we can filter the spectral components for $|w| > \pi/T_s$ in $X_c(w)$ before sampling operation. In this way, we can eliminate the overlapping shaded parts in Fig. 2.97. We name this filter as anti-aliasing filter and it is mathematically defined as

Fig. 2.94 Continuous to digital conversion

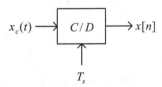

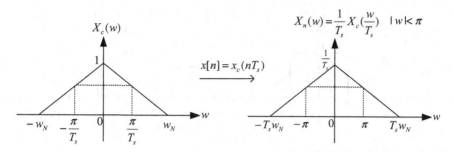

Fig. 2.95 Aliasing case in the Fourier transform of $x[n]$

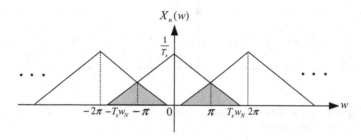

Fig. 2.96 Aliasing in $X_n(w)$

Fig. 2.97 $X_n(w)$ in Fig. 2.96
for $|w| < \pi$

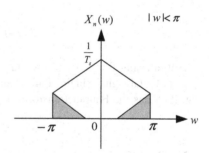

$$H_{aa}(w) = \begin{cases} 1 & if \quad |w| < \frac{\pi}{T_s} \\ 0 & otherwise \end{cases} \tag{2.107}$$

whose time domain expression can be computed using inverse Fourier transform

$$h_{aa}(t) = \frac{1}{2\pi} \int\limits_{-\infty}^{\infty} H_{aa}(w)e^{jwt}dw$$

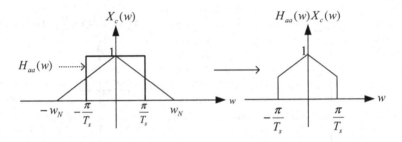

Fig. 2.98 Anti-aliasing filtering

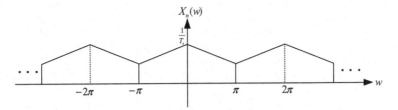

Fig. 2.99 The Fourier transform of a digital signal obtained by sampling of a continuous time signal filtered by an anti-aliasing filter

as

$$
h_{aa}(t) = \frac{\sin\left(\frac{\pi t}{T_s}\right)}{\pi t}.
\tag{2.108}
$$

Anti-aliasing filtering is shown in Fig. 2.98.

The digital signal obtained after sampling of the filtered analog signal shown in Fig. 2.98 has the Fourier transform depicted in Fig. 2.99.

2.3   Practical Implementations of C/D and D/C Converters

Up to now we have studied theoretical C/D and D/C converter systems. However, the practical implementation of these units in real life shows some differences. The practical implementation of the C/D converter is shown in the first part of Fig. 2.100, and in a similar manner, the practical implementation of the D/C converter is shown in the second part of Fig. 2.100.

C/D and D/C conversion systems include analog-to-digital and digital-to-analog converter units and the contents of these units are shown in Fig. 2.100. Now we will inspect every component of the complete system shown in Fig. 2.100.

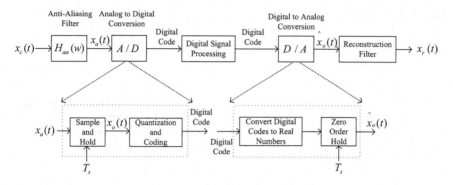

Fig. 2.100   Practical implementations of C/D and D/C converter systems

2.3.1   C/D Conversion

A practical C/D converter includes the units shown in Fig. 2.101.

Where antialiasing filter is used to decrease of amount of distortion in digital signal in case of aliasing. Antialiasing filter is defined as

$$H_{aa} = \begin{cases} 1 & |w| < \frac{\pi}{T_s} \\ 0 & otherwise \end{cases} \tag{2.109}$$

Inside A/D converter, we have Sample-and-Hold and Quantizer-Coder units which are shown in Fig. 2.102.

For the coding of quantization levels, two's complement, one's complement or unsigned binary representations can be used.

Once the analog signal is represented by bit sequences, i.e., codes, these bit sequences are processed depending on the application. For instance, in digital communication, these bit sequences are encoded by channel codes and obtained bit sequences are converted to complex symbols, i.e., digitally modulated, and transmitted. In data storage, these bit sequences are again coded using forward error corrections codes, such as Reed Solomon codes as in compact disc storage, and stored. Alternatively, these bit sequences can be passed through data compression algorithms and then stored.

Fig. 2.101   Practical C/D
converter.

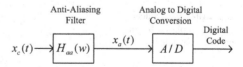

Fig. 2.102 Components of A/D converter

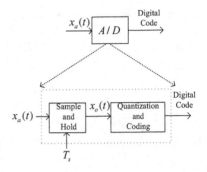

2.3.2 Sample and Hold

The aim of the sample and hold circuit is to produce a rectangular signal and the amplitudes of the rectangles are determined at the sampling time instants. The simplest sample and hold circuit as shown in Fig. 2.103 which is constructed using a capacitor.

Since usually sampling frequency f_s is a large number, such as 10 kHz etc., it is logical to use a digital switch for the place of a mechanical switch as shown in Fig. 2.104.

In the literature, much better sample and hold circuits are available. To give an idea about design improvement, the circuit in Fig. 2.104 can be improved by appending a buffer to the output preventing back current flows etc., and this improved circuit is shown in Fig. 2.105.

The sample and hold operation for the input sine signal is illustrated in Fig. 2.106.

Fig. 2.103 A simple sample and hold circuit

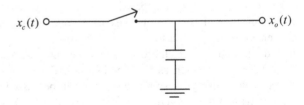

Fig. 2.104 Mechanical switch is replaced by an electronic switch

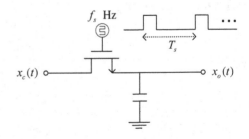

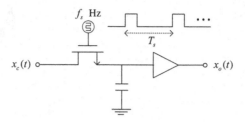

Fig. 2.105 Sample and hold circuit with a buffer at its output

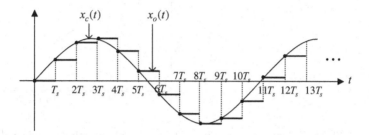

Fig. 2.106 Calculation of the output of the sample and hold circuit for sine input signal

For sine input signal after sample and hold operation, we obtain the signal $x_o(t)$ which is depicted alone in Fig. 2.107.

Question: Can we write a mathematical expression for the signal $x_o(t)$ shown in Fig. 2.107.

Yes, we can write. For this purpose, let's first define $h_o(t)$ function as shown in Fig. 2.108.

If the graph of $x_o(t)$ in Fig. 2.107 is inspected, it is seen that $x_o(t)$ signal is nothing but sum of the shifted and scaled $h_o(t)$ functions. Using $h_o(t)$ functions, we can write $x_o(t)$ as

$$x_o(t) = \sum_{k=-\infty}^{\infty} x_c(nT_s)h_o(t - nT_s) \rightarrow x_o(t) = \sum_{k=-\infty}^{\infty} x[n]h_o(t - nT_s). \qquad (2.110)$$

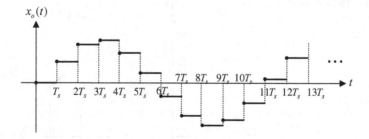

Fig. 2.107 Output of the sample and hold circuit for sine input signal

Fig. 2.108 Rectangle pulse signal

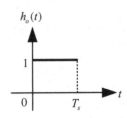

Fig. 2.109 Continuous time signal for sample and hold circuit

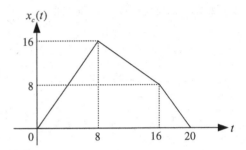

Example 2.30 The signal shown in Fig. 2.109 is passed through a sample and hold circuit. Find the signal at the output of the sample and hold circuit. Take sampling period as $T_s = 2$.

Solution 2.30 First we determine the amplitude values for the time instants t such that $t = nT_s$ where $T_s = 2$ and n is integer. This operation result is shown in Fig. 2.110. In addition, we also write the line equations for the computation of the amplitude values for the given time instants.

The amplitude values of the continuous time signal at time instants $t = nT_s$ are shown clearly in Fig. 2.111.

In the next step, we draw horizontal lines for the determined amplitudes, and for the first two samples, the drawn horizontal lines are shown in Fig. 2.112.

And for the first 4 samples, the horizontal drawn lines are shown in Fig. 2.113.

Repeating this procedure for all the other samples, we obtain the graph shown in Fig. 2.114.

The drawn horizontal lines for all the samples are depicted alone in Fig. 2.115.

Fig. 2.110 The continuous time signal in details

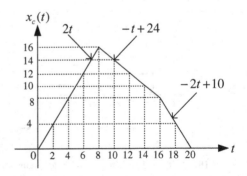

Fig. 2.111 Amplitudes shown explicitly for the time instants $t = nT_s$ where $T_s = 2$

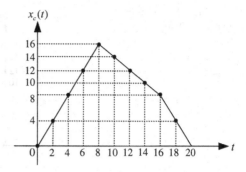

Fig. 2.112 Horizontal lines are drawn for the first two samples

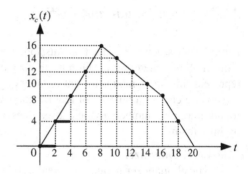

Fig. 2.113 Horizontal lines are drawn for the first four samples

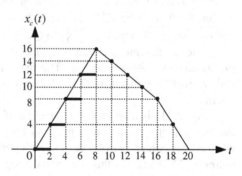

Fig. 2.114 Horizontal lines are drawn for all the samples

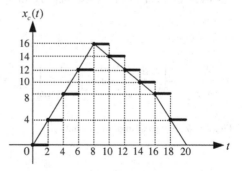

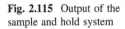

Fig. 2.115 Output of the
sample and hold system

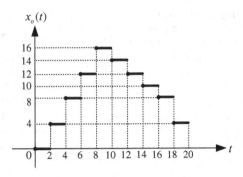

2.3.3   Quantization and Coding

During data storage or data transmission, we use bit sequences to represent real
number. Since there are an infinite number of real numbers, it is not possible to
represent this vast amount of real numbers by limited length bit streams. For this
reason, we choose a number of real numbers to represent by bit streams and try to
round other real numbers to the chosen ones when it is necessary to represent them
by bit streams.

Mid-Level Quantizer

A typical quantizer includes the real number intervals used to map real numbers
falling into these intervals to the quantization levels as shown in Fig. 2.116.

The quantizer in Fig. 2.116 is called mid-level quantizer. The quantizer maps the
real numbers in the range $\left[-\frac{\Delta}{2}, \frac{\Delta}{2}\right)$ to Q_0, maps the real numbers in the range $\left[\frac{\Delta}{2}, \frac{3\Delta}{2}\right)$
to Q_1 etc. In this quantizer, Δ is called the step size of the quantizer. Smaller Δ
means more sensitive quantizer. The mapping between real numbers and quantization levels is defined as $Q_i = Q(x)$ where Q_i may be chosen as the center of
interleaves.

If Fig. 2.116 is inspected, it is seen that if we have equal number of intervals on
the negative and positive regions, it means that the total number of intervals is an
odd number, which is not a desired situation. Since using N bits, it is possible to
represent 2^N levels. For this reason, we design these quantizers such that if one side
has even number of intervals, then the other side has odd number of intervals.

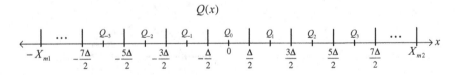

Fig. 2.116 A typical mid-level quantizer

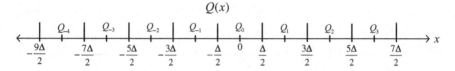

Fig. 2.117 Mid-level quantizer for Example 2.31

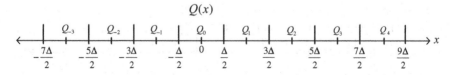

Fig. 2.118 An alternative mid-level quantizer for Example 2.31

Example 2.31 A 3-bit quantizer includes $2^3 = 8$ quantization intervals. A mid-level type quantizer consisting of 8 levels can be shown as in Fig. 2.117.

Or alternatively as in Fig. 2.118.

We will use mid-level quantizers as in Fig. 2.117.

As it is clear from the Example 2.30, for an N-bit mid-level quantizer, the minimum number that can be quantized is $-(2^N + 1)/2$ and the maximum number that can be quantized is $(2^N - 1)/2$.

The quantization levels are represented by binary sequences, such as two's complement, one's complements, unsigned representation, or private bit sequences can be assigned for quantization levels.

Example 2.32 Design a 3-bit quantizer for the real numbers in the range $[-14 \cdots 14]$.

Solution 2.32 For a 3-bit quantizer $X_{m1} = -9\Delta/2$ and $X_{m2} = 7\Delta/2$. Equating X_{m2} to -14, we obtain

$$\frac{7\Delta}{2} = 14 \rightarrow \Delta = 4.$$

So our quantizer can quantize the real numbers in the range

$$\left[-\frac{9\Delta}{2} \cdots \frac{7\Delta}{2}\right] = [-18 \cdots 14].$$

The bit sequences for our quantizer can be assigned to the intervals as in Fig. 2.119 and centers of the interleavers can be calculated as in Fig. 2.120.

Mid-Rise Quantizer

The mid-rise quantizer is shown in Fig. 2.121. As it is clear from Fig. 2.121, there is no interval centered at the origin.

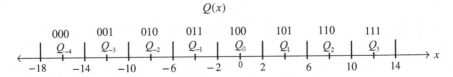

Fig. 2.119 Bit sequences assigned to the quantization intervals

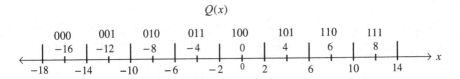

Fig. 2.120 Mid-level quantizer

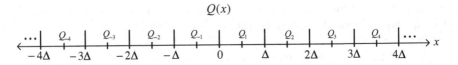

Fig. 2.121 Mid-rise quantizer

Assume that we want to quantize a sequence of digital samples represented by $x[n]$. Let $\hat{x}[n]$ be the sequence obtained after quantization. Since quantization distorts the original signal, the quantized samples mathematically can be written as

$$\hat{x}[n] = Q(x[n]) \rightarrow \hat{x}[n] = x[n] + e[n] \qquad (2.111)$$

where $e[n]$ is called quantization noise.

2.3.4   D/C Converter

The practical implementation of D/C converter is shown in Fig. 2.122.

The content of the D/A converter is detailed in Fig. 2.123.

The digital codes are converted to real numbers according to the used coding scheme. At the output of the code-to-digital converter, we have digital samples which can be written as

Fig. 2.122 D/C conversion

Fig. 2.123 D/A conversion

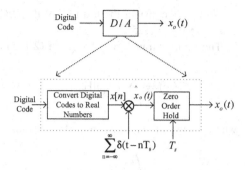

Fig. 2.124 Impulse response of zero order hold

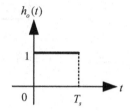

$$\hat{x}[n] = x[n] + e[n] \tag{2.112}$$

where $e[n]$ is the quantization error. The zero order hold filter impulse response is shown Fig. 2.124.

The output of the code-to-digital converter in Fig. 2.123 is

$$\hat{x}_o(t) = \sum_{n=-\infty}^{\infty} \hat{x}[n]\delta(t - nT_s). \tag{2.113}$$

When $\hat{x}_o(t)$ is passed through zero order hold filter, we obtain

$$x_o(t) = \hat{x}_o(t) * h_o(t) \rightarrow x_o(t) = \sum_{n=-\infty}^{\infty} \hat{x}[n]h_o(t - nT_s). \tag{2.114}$$

Substituting $\hat{x}[n] = x[n] + e[n]$ in (2.114), we get

$$x_o(t) = \sum_{n=-\infty}^{\infty} x[n]h_o(t - nT_s) + \sum_{n=-\infty}^{\infty} e[n]h_o(t - nT_s). \tag{2.115}$$

Fig. 2.125 Reconstruction filter block diagram

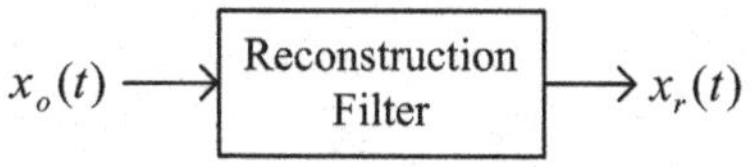

$$x_o(t) \longrightarrow \boxed{\begin{array}{c} \text{Reconstruction} \\ \text{Filter} \end{array}} \longrightarrow x_r(t)$$

Now let's consider the last unit of the D/C converter the reconstruction filter as shown in Fig. 2.125.

The Fourier transform of $x_o(t)$ in (2.115) can be calculated using

$$X_o(w) = \sum_{n=-\infty}^{\infty} x[n] H_o(w) e^{-jwnT_s} + \sum_{n=-\infty}^{\infty} e[n] E_o(w) e^{-jwnT_s} \qquad (2.116)$$

where taking the common term $H_o(w)$ outside the parenthesis, we obtain

$$X_o(w) = \left(\underbrace{\sum_{n=-\infty}^{\infty} x[n] e^{-jwnT_s}}_{X_n(T_s w)} + \underbrace{\sum_{n=-\infty}^{\infty} e[n] e^{-jwnT_s}}_{E_n(T_s w)} \right) H_o(w) \qquad (2.117)$$

which can be written as

$$X_o(w) = (X_n(T_s w) + E_n(T_s w)) H_o(w). \qquad (2.118)$$

From Fig. 2.125, we can write

$$X_r(w) = H_r(w) X_o(w) \qquad (2.119)$$

where $H_r(w)$ is the frequency response of the reconstruction filter. If we choose $H_r(w)$ as

$$H_r(w) = \begin{cases} \frac{T_s}{H_o(w)} & |w| < \frac{\pi}{T_s} \\ 0 & otherwise \end{cases} \qquad (2.120)$$

and substituting it into (2.119) and using (2.118) in (2.119), we obtain

$$X_r(w) = T_s X_n(T_s w) + T_s E_n(T_s w) \quad |w| < \frac{\pi}{T_s} \qquad (2.121)$$

which is the Fourier transform of

$$x_r(t) = x_a(t) + e(t). \qquad (2.122)$$

Since $x[n] = x_a(nT_s)$, $e[n] = e(nT_s)$, the continuous time signals $x_a(t)$ and $e(t)$ can be obtained from their samples using

$$x_a(t) = \sum_{n=-\infty}^{\infty} x[n] \sin c\left(\frac{t - nT_s}{T_s}\right) \qquad (2.123)$$

and

$$e(t) = \sum_{n=-\infty}^{\infty} e[n] \sin c\left(\frac{t - nT_s}{T_s}\right). \qquad (2.124)$$

Then $x_r(t)$ in (2.122) using (2.123) and (2.124) can be written as

$$x_r(t) = \sum_{n=-\infty}^{\infty} x[n] \sin c\left(\frac{t - nT_s}{T_s}\right) + \sum_{n=-\infty}^{\infty} e[n] \sin c\left(\frac{t - nT_s}{T_s}\right).$$

2.4   Problems

(1) $x[n] = \begin{bmatrix} 1 & 2 & 0 & -3 & -1 & 1 & 4 & -1 & 0 & 1 & -2 & 5 & 1 & 3 \end{bmatrix}$ is
given. Find the signals $x[2n]$, $x[3n]$, $x[4n]$, $x[n/2]$, $x[n/3]$, and $x[n/4]$.
(2) One period of the Fourier transform of $x[n]$ around origin is shown in
Fig. 2.126. Draw the Fourier transform of the downsampled signal
$y[n] = x[2n]$.
(3) The delay system is described in Fig. 2.127.

Fig. 2.126 One period of
$X_n(w)$ around origin

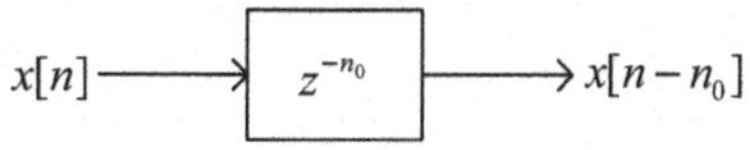

Fig. 2.127 Delay system

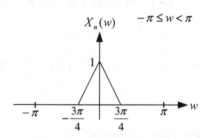

Fig. 2.128 Signal processing
system

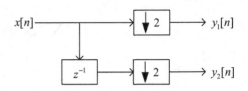

If

$$x[n] = [a \quad b \quad c \quad d \quad e \quad f \quad \underset{n=0}{g} \quad h \quad \imath \quad j \quad k \quad l \quad m \quad n \quad o \quad p \quad r]$$

find the output of each unit in Fig. 2.128.

(4) Calculate the inverse Fourier transform of the digital filter

$$H_{dn}(w) = \begin{cases} 1 & if \quad |w| < \frac{\pi}{M} \\ 0 & if \quad \frac{\pi}{M} < |w| < \pi. \end{cases} \tag{2.125}$$

(5) Draw the graph of

$$h_{dn}[n] = \frac{\sin\left(\frac{\pi n}{M}\right)}{\pi n} \tag{2.126}$$

roughly, and find the triangle approximation of (2.126). Calculate the approximated
model for $n = -5, \ldots, 5$.

(6) The graph of $X(t)$ is shown in Fig. 2.129. Considering Fig. 2.129 draw the
graph of

$$Y(t) = \sum_{k=-\infty}^{\infty} X(t - kT), \quad T = 3. \tag{2.127}$$

Fig. 2.129 The graph of $X(t)$

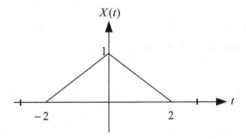

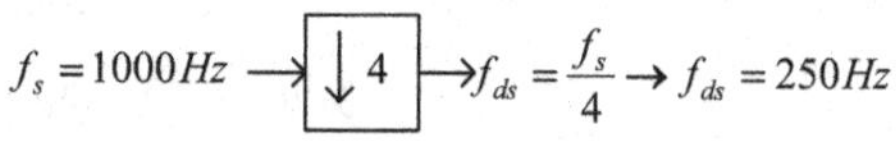

Fig. 2.130 Downsampler

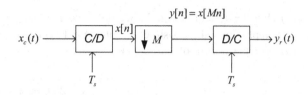

$$f_s = 1000\,Hz \longrightarrow \boxed{\downarrow 4} \longrightarrow f_{ds} = \frac{f_s}{4} \longrightarrow f_{ds} = 250\,Hz$$

Fig. 2.131 System for
Question 9

$$y[n] = x[Mn]$$

$$x_c(t) \longrightarrow \boxed{C/D} \xrightarrow{x[n]} \boxed{\downarrow M} \longrightarrow \boxed{D/C} \longrightarrow y_r(t)$$

$$\qquad\qquad\uparrow\qquad\qquad\qquad\qquad\uparrow$$
$$\qquad\qquad T_s\qquad\qquad\qquad\qquad T_s$$

(7) Repeat Question-6 for $T = 1$, $T = 4$ and $T = 5$.

(8) Comment on the system shown in Fig. 2.130.

(9) For the system of Fig. 2.131, $x_c(t)$ is a lowpass signal with bandwidth 3000 Hz, $T_s = \frac{1}{8000}$ s and $M = 2$. Is system output $y_r(t)$ equal to system input $x_c(t)$? If they are equal to each other, justify the reasoning behind it. If they are not equal to each other, again explain the reasoning behind it.

(10) If $x[n] = [1 \quad 2 \quad 3 \quad 4 \quad 5 \quad 6 \quad 7]$ and $L = 4$, draw the graph of

$$y[n] = \sum_{k=-\infty}^{\infty} x[k]\delta[n - kL].$$

(11) For the system of Fig. 2.132, $M = L = 2$ and

$$x[n] = [a \quad b \quad c \quad d \quad e \quad f \quad g \quad h \quad \underbrace{l}_{n=0} \quad j \quad k \quad l \quad m \quad n \quad o \quad p \quad r \quad s].$$

Find $x_d[n]$ and $y[n]$.

(12) Draw the graph of $h_{ai}[n] = -\frac{|n|}{L} + 1$, $-L \leq n \leq L$ for $L = 3$ and $L = 8$.

(13) $x_d[k] = [1 \quad 4 \quad 7 \quad 10 \quad 13]$, $h_{ai}[n] = -\frac{|n|}{L} + 1$, $-L \leq n \leq L$, $L = 3$, calculate and draw

Fig. 2.132 Signal processing
system

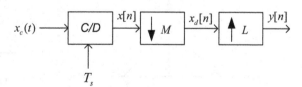

$$x_c(t) \longrightarrow \boxed{C/D} \xrightarrow{x[n]} \boxed{\downarrow M} \xrightarrow{x_d[n]} \boxed{\uparrow L} \xrightarrow{y[n]}$$

$$\qquad\qquad\uparrow$$
$$\qquad\qquad T_s$$

$$y_i[n] = \sum_{k=-\infty}^{\infty} x_d[k]h_{ai}[n - kL].$$

(14) For the system of Fig. 2.133, $x[n] = \cos\left(\frac{\pi}{4}n\right)$ $0 \le n \le 10$, $h_{ai}[n]$ is the triangle approximated reconstruction filter. Find $x_d[n], y[n]$ and $y_i[n]$ for $M = L = 2$.

(15) For the system of Fig. 2.134,

$$H_{aa}(w) = \begin{cases} 1 & if\ |w| < \frac{\pi}{T_s} \\ 0 & otherwise \end{cases}$$

Express the Fourier transform of $x[n]$ in terms of the Fourier transform of $x_c(t)$.

(16) For the system of Fig. 2.135, $M = 3$, $X_n(w)$ is the one period of the Fourier transform of $x[n]$. Draw the Fourier transform of $x_d[n]$.

(17) For the system of Fig. 2.136, $M = L = 2$ and $X_n(w)$ is the one period of the Fourier transform of $x[n]$.

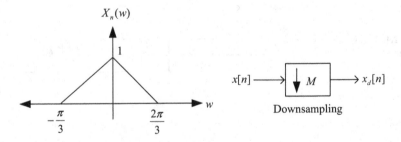

Fig. 2.133 Signal processing system for Question 14

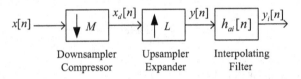

Fig. 2.134 Signal processing system for Question 15

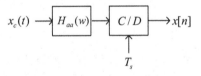

Fig. 2.135 Downsampling of digital signal

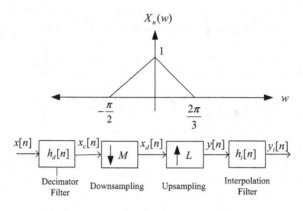

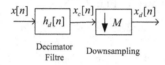

Fig. 2.136 Signal processing system for Question 17

Fig. 2.137 Decimation system

(a) Draw the Fourier transforms of $x_c[n], x_d[n]$, and $y[n]$.
(b) Draw the triangle approximation model of the interpolation filter for $L = 2$.
(c) Draw the Fourier transform of $y_i[n]$ for $\sin c(\cdot)$ interpolation filter.
(d) If $x_c[n] = [1.0 \quad 1.7 \quad 2.4 \quad 3.2 \quad 4]$, calculate $x_d[n], y[n]$, using triangle approximated interpolation filter.

(18) For the system of Fig. 2.137, $M = 2$, and $H_d(w)$ is defined as

$$H_d(w) = \begin{cases} 1 & if \quad |w| \leq \frac{\pi}{M} \\ 0 & otherwise \end{cases} \tag{2.128}$$

(a) Calculate the inverse Fourier transform of $H_d(w)$, i.e., calculate $h_d[n]$. Next, find the triangle approximated model of $h_d[n]$.
(b) For $x[n] = [1 \quad 2 \quad 3 \quad 4]$ calculate $x_o[n]$ and $x_d[n]$.

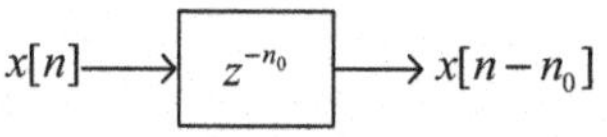

$$x[n] \longrightarrow \boxed{z^{-n_0}} \longrightarrow x[n-n_0]$$

Fig. 2.138 Delay system

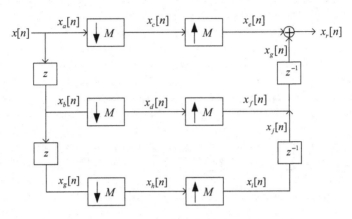

Fig. 2.139 Signal processing system for Question 19

(19) The delay system is shown in Fig. 2.138.

For the system of Fig. 2.139, $M = 3$, $x[n] = [a \quad b \quad c \quad d \quad e \quad f \quad g$ $h \quad i \quad j \quad k \quad l \quad m \quad n \quad o \quad p \quad r \quad s \quad t \quad u \quad v \quad w \quad x \quad y]$. Find the signal at the output of each unit, and find the system output.

Chapter 3
Discrete Fourier Transform

In linear algebra, basis vectors span the entire vector space. And any vector of the vector space can be written as the linear combination of the basis vectors. For any vector in vector space, finding the coefficients of basis vectors used for the construction of the vector can be considered as a transformation. Fourier series are used to represent periodic signals. Fourier series are used to construct any periodic signal from sinusoidal signals. The sinusoidal signals can be considered as the basis signals, and linear combination of these signals with complex coefficients produce any periodic signal. Once we obtain the coefficients of the base signals necessary for the construction of a periodic signal, then we have full knowledge of the periodic signal and instead of transmitting the periodic signal, we can transmit the coefficients of the base signals. Since at the receiver side, the periodic signal can be reconstructed using the base coefficients.

In this chapter we will study a new transformation technique called discrete Fourier transform used for aperiodic digital signals. We will show that similar to the Fourier series representation of periodic digital signals, aperiodic digital signals can also be written as a linear combination of sinusoidal digital aperiodic signals. In this case, aperiodic digital sinusoidal signals can be considered as base signals. And finding the coefficients of base signals such that their linear combination yields the aperiodic digital signal is called discrete Fourier transformation of the aperiodic digital signal. Thus, the discrete Fourier transformation is nothing but finding the set of coefficients of the base signals for an aperiodic digital signal. And once we have these coefficients, then we have full knowledge on the aperiodic signal in another digital sequence.

© Springer Nature Singapore Pte Ltd. 2018
O. Gazi, *Understanding Digital Signal Processing*, Springer Topics in Signal Processing 13, DOI 10.1007/978-981-10-4962-0_3

3.1 Manipulation of Digital Signals

Before studying discrete Fourier transform, let's prepare ourselves for the subject, for this purpose, we will first study the manipulation of digital signals.

Manipulation of Non-periodic Digital Signals

A non-periodic or aperiodic digital signal has finite number of samples. And these signals are illustrated either by graphics or by number vectors, or by number sequences. As an example, a digital signal and its vector representation is shown in Fig. 3.1.

Manipulation of digital signals includes shifting, scaling in time domain and change in amplitudes.

Shifting of Digital Signals in Time Domain

Given $x[n]$, to obtain $x[n - n_0], n_0 > 0$, we shift the amplitudes of $x[n]$ to the right by n_0 units. If $n_0 < 0$, amplitudes are shifted to the left.

Shifting amplitudes to the right by n_0 equals to the shifting $n = 0$ index to the left by n_0 units. This operation is illustrated in the following example.

Example 3.1 Given
$$x[n] = [a \quad b \quad c \quad d \quad \underbrace{e}_{n=0} \quad f \quad g \quad h \quad i \quad j \quad k], \text{ find } x[n - 1]$$
$x[n - 3], \quad x[n + 1], \quad x[n + 2], \text{ and } x[n - 7].$

Solution 3.1 To get $x[n - 1]$, we shift amplitudes of $x[n]$ to the right by '1' unit. Shifting amplitudes to the right by '1' unit is the same as shifting $n = 0$ index to the left by '1' unit, the result of this operation is

$$x[n - 1] = [a \quad b \quad c \quad \underbrace{d}_{n=0} \quad e \quad f \quad g \quad h \quad i \quad j \quad k].$$

Following a similar approach for $x[n - 3]$, we obtain

$$x[n - 3] = [a \quad \underbrace{b}_{n=0} \quad c \quad d \quad e \quad f \quad g \quad h \quad i \quad j \quad k].$$

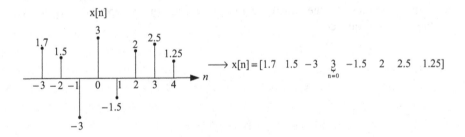

Fig. 3.1 A digital signal and its representation by a number vector

To get $x[n+1]$, we shift amplitudes of $x[n]$ to the left by '1' unit obtaining

$$x[n+1] = [a \quad b \quad c \quad d \quad e \quad \underbrace{f}_{n=0} \quad g \quad h \quad i \quad j \quad k].$$

Following similar steps, we obtain

$$x[n+2] = [a \quad b \quad c \quad d \quad e \quad f \quad \underbrace{g}_{n=0} \quad h \quad i \quad j \quad k],$$

and

$$x[n-7] = [\underbrace{0}_{n=0} \quad 0 \quad 0 \quad a \quad b \quad c \quad d \quad e \quad f \quad g \quad h \quad i \quad j \quad k].$$

where it is clear that if the shifting amount goes beyond the signal frontiers, for the new time instants, 0 values are assigned for the signal amplitudes.

Scaling of Digital Signals in Time Domain

To find $x[Mn]$, we divide the time axis of $x[n]$ by M, and keep only integer division results and omit the non-integer division results. The resulting signal is nothing but $x[Mn]$.

Example 3.2 If $x[n] = [a \quad b \quad c \quad d \quad \underbrace{e}_{n=0} \quad f \quad g \quad h \quad i \quad j \quad k]$, find $x[2n]$ and $x[3n]$.

Solution 3.2 To get $x[2n]$, we divide time axis of $x[n]$ by 2 and keep only integer division results. First, let's write all the time indices as shown in

$$[\underbrace{a}_{-4} \quad \underbrace{b}_{-3} \quad \underbrace{c}_{-2} \quad \underbrace{d}_{-1} \quad \underbrace{e}_{n=0} \quad \underbrace{f}_{1} \quad \underbrace{g}_{2} \quad \underbrace{h}_{3} \quad \underbrace{i}_{4} \quad \underbrace{j}_{5} \quad \underbrace{k}_{6}].$$

$$(3.1)$$

Next, we divide the indices as in

$$[\underbrace{a}_{-\frac{4}{2}} \quad \underbrace{b}_{-\frac{3}{2}} \quad \underbrace{c}_{-\frac{2}{2}} \quad \underbrace{d}_{-\frac{1}{2}} \quad \underbrace{e}_{\frac{0}{2}} \quad \underbrace{f}_{\frac{1}{2}} \quad \underbrace{g}_{\frac{2}{2}} \quad \underbrace{h}_{\frac{3}{2}} \quad \underbrace{i}_{\frac{4}{2}} \quad \underbrace{j}_{\frac{5}{2}} \quad \underbrace{k}_{\frac{6}{2}}]$$

$$(3.2)$$

where keeping only integer division results, we obtain

$$x[2n] = [\underbrace{a}_{-2} \quad \underbrace{b}_{-2} \quad \underbrace{c}_{0} \quad \underbrace{g}_{1} \quad \underbrace{i}_{2} \quad \underbrace{k}_{3}]$$

which can be written in its simple form as

$$x[2n] = [a \quad c \quad \underbrace{e}_{0} \quad g \quad i \quad k].$$

Following a similar approach for $x[3n]$, we obtain

$$x[3n] = [b \quad \underbrace{e}_{n=0} \quad h \quad k].$$

Combined Shifting and Scaling
To obtain $x[Mn - n_0]$, we follow a two-step procedure as listed below.

(1) First, the shifted signal, $x[n - n_0]$ is obtained, and this signal is denoted by $x_1[n]$, i.e., $x_1[n] = x[n - n_0]$

(2) Then using $x_1[n]$, we obtain the scaled signal $x_1[Mn]$ which is nothing but $x[Mn - n_0]$

That is, we first obtain the shifted signal $x_1[n] = x[n - n_0]$, and then using $x_1[n]$ we get the scaled signal $x_1[Mn] = x[Mn - n_0]$.

Example 3.3 If $x[n] = [a \quad b \quad c \quad d \quad \underbrace{e}_{n=0} \quad f \quad g \quad h \quad i \quad j \quad k]$, find $x[3n+3]$.

Solution 3.3 First, we obtain the shifted signal $x[n+3]$ as

$$x[n+3] = [a \quad b \quad c \quad d \quad e \quad f \quad g \quad \underbrace{h}_{n=0} \quad i \quad j \quad k].$$

Let $x_1[n] = x[n+3]$, i.e., $x_1[n] = [a \quad b \quad c \quad d \quad e \quad f \quad g \quad \underbrace{h}_{n=0} \quad i \quad j \quad k]$, then the scaled signal $x_1[3n]$ can be calculated as

$$x_1[3n] = [b \quad e \quad \underbrace{h}_{n=0} \quad k]$$

which is nothing but $x[3n+3]$, that is

$$x[3n+3] = [b \quad e \quad \underbrace{h}_{n=0} \quad k].$$

Note: If $n = 0$ index is not indicated in the digital signal vector representation, then the first element index is accepted as $n = 0$.

3.1.1 Manipulation of Periodic Digital Signals

Manipulation of periodic digital signals includes shifting, scaling and combined shifting, scaling operations. There is no difference in manipulating non-periodic and periodic digital signals. The same set of operations are applied for the manipulation of periodic signals as in the manipulation of non-periodic signals.

However, since periodic signals are of infinite lengths, for easy of manipulation, it is logical to consider just one period of the periodic signal and perform manipulations on it.

Let $\tilde{x}[n]$ be a periodic signal with fundamental period N, i.e., $\tilde{x}[n] = \tilde{x}[n + lN]$ $l, N \in Z$.

Let's define one period of $\tilde{x}[n]$ as

$$x[n] = \begin{cases} \tilde{x}[n] & 0 \leq n \leq N - 1 \\ 0 & otherwise. \end{cases} \tag{3.3}$$

Using (3.3), we can write $\tilde{x}[n]$ in terms of $x[n]$ as

$$\tilde{x}[n] = \sum_{k=-\infty}^{\infty} x[n - kN]. \tag{3.4}$$

3.1.2 Shifting of Periodic Digital Signals

First let's make definitions as follows:
Rotate Right
When the signal $x[n] = [\,1 \quad 2 \quad 3 \quad 4 \quad \cdots \quad N\,]$ is rotated right, we get

$$RR(x[n]) = [\,N \quad 1 \quad 2 \quad 3 \quad 4 \quad \cdots \quad N - 1\,]. \tag{3.5}$$

$RR(x[n], m)$ is the m unit rotated (right) signal.
Rotate Left
When the signal $x[n] = [\,1 \quad 2 \quad 3 \quad 4 \quad \cdots \quad N\,]$ is rotated left, we get

$$RL(x[n]) = [\,2 \quad 3 \quad 4 \quad \cdots \quad N - 1 \quad N \quad 1\,]. \tag{3.6}$$

$RL(x[n], m)$ is the m unit rotated (left) signal.
Rotate Inside
When the signal $x[n] = [\,1 \quad 2 \quad 3 \quad 4 \quad \cdots \quad N\,]$ is rotated inside, we get

$$RI(x[n]) = \begin{bmatrix} 1 & N & N-1 & N-2 & \cdots & 2 \end{bmatrix}. \tag{3.7}$$

Shifting of Periodic Digital Signals

If $x[n]$ is the one period of the periodic signal, $\tilde{x}[n]$ such that $0 \le n \le N-1$, one period of the shifted signal $\tilde{x}[n-n_0]$, $n_0 > 0$ is obtained by rotating amplitudes of $x[n]$ to the right (left if $n_0 < 0$) by n_0 units.

Example 3.4 The signal given in Fig. 3.2 is a periodic signal, i.e., $\tilde{x}[n] = \tilde{x}[n+N]$. Find the period of this signal, and determine its one period for $0 \le n \le N-1$.

Solution 3.4 To find the period of the signal, we need to find the repeating pattern in the signal graph. If the signal shown in Fig. 3.2 is carefully inspected the repeating pattern can be easily determined. The repeating pattern of Fig. 3.2 is shown in Fig. 3.3 in bold. The number of samples in the repeating pattern is nothing but the period of the signal. Hence, for this example, N the period of the signal is 5, i.e., $\tilde{x}[n] = \tilde{x}[n+5]$.

One period of the signal in Fig. 3.3 for $0 \le n \le 4$ is shown in Fig. 3.4.

Using one period of the signal starting at origin, we can write the periodic signal as

$$\tilde{x}[n] = [\cdots \quad \underbrace{3}_{n=0} \quad -1.5 \quad 1.7 \quad 1.5 \quad -3 \quad \cdots].$$

Fig. 3.2 A periodic digital signal

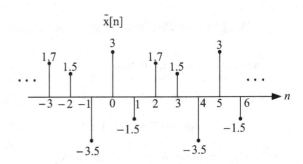

Fig. 3.3 The repeating pattern of Fig. 3.2 is shown in bold

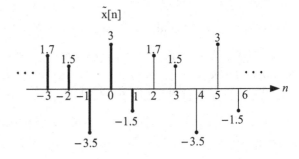

Fig. 3.4 One period of the signal in Fig. 3.3 for $0 \leq n \leq 4$

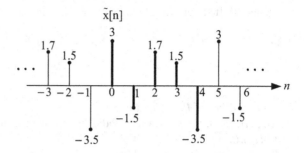

Fig. 3.5 The periodic signal $\tilde{x}[n]$ for Example 3.5

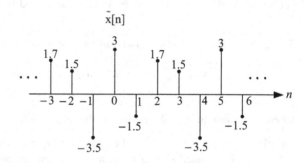

Example 3.5 The periodic signal $\tilde{x}[n]$ is shown in Fig. 3.5, find $\tilde{x}[n - 3]$, and $\tilde{x}[n + 2]$.

Solution 3.5 The period of the signal is $N = 5$, and signal amplitudes for one period are

$$x[n] = [\underbrace{3}_{n=0} \quad -1.5 \quad 1.7 \quad 1.5 \quad -3.5]. \tag{3.8}$$

When $x[n]$ is rotated to the right by 3 units, we get

$$RR(x[n], 3) = [\underbrace{1.7}_{n=0} \quad 1.5 \quad -3.5 \quad 3 \quad -1.5]. \tag{3.9}$$

And using (3.9), we can write the shifted periodic signal as

$$\tilde{x}[n - 3] = [\cdots \quad 1.7 \quad 1.5 \quad -3.5 \quad 3 \quad -1.5 \quad \underbrace{1.7}_{n=0} \quad 1.5 \quad -3.5 \quad 3 \quad -1.5 \quad \cdots].$$

$$\tag{3.10}$$

To find $\tilde{x}[n + 2]$, one period of $\tilde{x}[n]$ is rotated to the left by 2 units yielding

$$RL(x[n], 2) = [\underbrace{1.7}_{n=0} \quad 1.5 \quad -3.5 \quad 3 \quad -1.5]. \tag{3.11}$$

Hence, shifted periodic signal $\tilde{x}[n+2]$ becomes as

$$\tilde{x}[n+2] = [\cdots 1.5 \quad -3.5 \quad 3 \quad -1.5 \quad \underset{n=0}{\underline{1.7}} \quad 1.5 \quad -3.5 \quad 3 \quad -1.5 \quad 1.7 \cdots].$$

$$(3.12)$$

Time Scaling of Periodic Signals

To perform time scaling on periodic signals, we consider one period of the signal and perform time scaling on it.

The resulting signal is nothing but the one period of the scaled signal. If the period of the digital signal $\tilde{x}[n]$ is N, then the period of the scaled signal $\tilde{x}[Mn]$ is N/M.

Example 3.6 The periodic signal $\tilde{x}[n]$ in its one interval equals to

$$x[n] = [\underset{n=0}{\underline{3}} \quad -1.5 \quad 1.7 \quad 1.5 \quad -3.5 \quad 2.2 \quad 4] \qquad (3.13)$$

where it is obvious that the period of the signal is $N = 7$. Find $\tilde{x}[2n]$ and $\tilde{x}[3n]$.

Solution 3.6 One period of $\tilde{x}[2n]$ equals to $x[2n]$, and one period of $\tilde{x}[3n]$ equals to $x[3n]$. The time scaled signals $x[2n]$ and $x[3n]$ can be calculated as

$$x[2n] = [\underset{n=0}{\underline{3}} \quad 1.7 \quad -3.5 \quad 4]$$

$$x[3n] = [\underset{n=0}{\underline{3}} \quad 1.5 \quad 4]. \qquad (3.14)$$

And using (3.14) the periodic signals $\tilde{x}[2n]$ and $\tilde{x}[3n]$ can be written as

$$\tilde{x}[2n] = [\cdots 1.7 \quad -3.5 \quad 4 \quad \underset{n=0}{\underline{3}} \quad 1.7 \quad -3.5 \quad 4 \quad 3 \quad 1.7 \quad -3.5 \quad 4 \cdots]$$

$$\tilde{x}[3n] = [\cdots 3 \quad 1.5 \quad 4 \quad \underset{n=0}{\underline{3}} \quad 1.5 \quad 4 \quad 3 \quad 1.5 \quad 4 \cdots].$$

Combined Shifting and Scaling

The periodic digital signal $\tilde{x}[n]$ can be shifted and scaled in time domain yielding the periodic signal $\tilde{x}[Mn - n_0]$. The shifted and scaled signal $\tilde{x}[Mn - n_0]$ can be obtained from $\tilde{x}[n]$ via a two-step procedure as explained below.

(1) To get $\tilde{x}[Mn - n_0]$, first the shifted signal $\tilde{x}[n - n_0]$ is obtained. Let's call this signal $\tilde{x}_1[n]$, i.e., $\tilde{x}_1[n] = \tilde{x}[n - n_0]$.
(2) In the next step, $\tilde{x}_1[n]$ is scaled in time domain and $\tilde{y}[n] = \tilde{x}_1[Mn]$ is obtained, and $\tilde{y}[n]$ is nothing but $\tilde{x}[Mn - n_0]$, i.e., $\tilde{y}[n] = \tilde{x}[Mn - n_0]$.

Example 3.7 The periodic signal $\tilde{x}[n]$ in its one interval equals to

$$x[n] = [\underbrace{3}_{n=0} \quad -1.5 \quad 1.7 \quad 1.5 \quad -3.5 \quad 2.2 \quad 4]$$

where it is obvious that the period of the signal is $N = 7$. Find $\tilde{x}[2n - 3]$.

Solution 3.7 To obtain $\tilde{x}[2n - 3]$, let's first find one period of the shifted signal $\tilde{x}[n - 3]$. One period of $\tilde{x}[n - 3]$ is obtained by rotating one period of $\tilde{x}[n]$ to the right by 3 yielding

$$RR(x[n], 3) = [\underbrace{-3.5}_{n=0} \quad 2.2 \quad 4 \quad 3 \quad -1.5 \quad 1.7 \quad 1.5] \qquad (3.15)$$

Let's denote (3.15) by $x_1[n]$, i.e., one period of $\tilde{x}[n] = \tilde{x}[n - 3]$, then we have

$$x_1[n] = [\underbrace{-3.5}_{n=0} \quad 2.2 \quad 4 \quad 3 \quad -1.5 \quad 1.7 \quad 1.5]. \qquad (3.16)$$

Next using (3.16), we can evaluate $x_1[2n]$ which is nothing but one period of $\tilde{x}[2n - 3]$ as

$$x_1[2n] = [\underbrace{-3.5}_{n=0} \quad 4 \quad -1.5 \quad 1.5].$$

Hence, our periodic signal $\tilde{x}[2n - 3]$ becomes as

$$\tilde{x}[2n - 3] = [\cdots -1.5 \quad 1.5 \quad \underbrace{-3.5}_{n=0} \quad 4 \quad -1.5 \quad 1.5 \quad -3.5 \quad 4 \quad -1.5 \cdots].$$

Example 3.8 Periodic signal $\tilde{x}[n]$ is shown in Fig. 3.6.
Find $\tilde{x}[-n]$.

Solution 3.8 To find $\tilde{x}[-n]$, we divide the time axis of $\tilde{x}[n]$ by -1. This operation is illustrated in Fig. 3.7.

The division operations in Fig. 3.7 yields the signal in Fig. 3.8.

When amplitudes and time indices are re-ordered together, we obtain the graph in Fig. 3.9.

Practical way to find $\tilde{x}[-n]$ signal

Fig. 3.6 Periodic signal $\tilde{x}[n]$ for Example 3.8

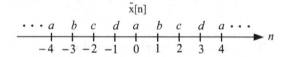

Fig. 3.7 Calculation of $\tilde{x}[-n]$

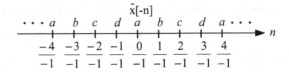

Fig. 3.8 After division of the time axis in Fig. 3.7

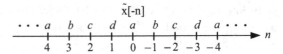

Fig. 3.9 Time axis re-ordered

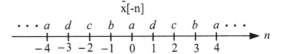

Fig. 3.10 Periodic signal $\tilde{x}[n]$

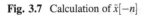

If one period of $\tilde{x}[n]$ is denoted by $x[n] = [\,1 \quad 2 \quad 3 \quad 4 \quad \cdots \quad N\,]$, then one period of $\tilde{x}[-n]$ can be obtained rotating $x[n]$ inside by 1 unit. That is, one period of $\tilde{x}[-n]$ is

$$RI(x[n]) = [\,1 \quad N \quad N-1 \quad N-2 \quad \cdots \quad 2\,] \tag{3.17}$$

We can apply this practical method to the previous example where the periodic signal had been given as in Fig. 3.10.

One period of is $\tilde{x}[n]$ in Fig. 3.10 is

$$x[n] = [a \quad b \quad c \quad d]. \tag{3.18}$$

When (3.18) is rotated inside, we obtain

$$RR(x[n]) = [a \quad b \quad c \quad d] \tag{3.19}$$

which is nothing but one period of $\tilde{x}[-n]$. Hence $\tilde{x}[-n]$ equals to

$$\tilde{x}[-n] = [\cdots d \quad c \quad b \underbrace{a}_{n=0} d \quad c \quad b \quad a \quad d \quad c \quad b \cdots].$$

Calculation of the periodic signal $\tilde{x}[n_0 - n]$

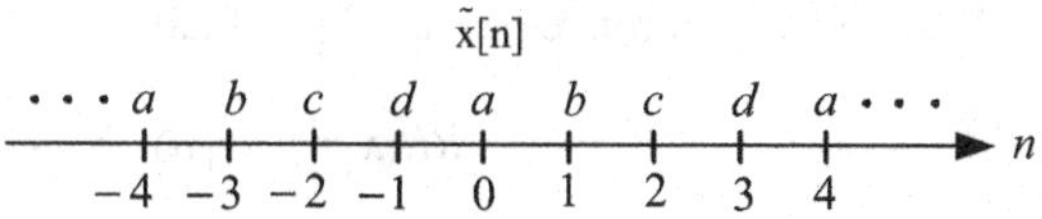

Fig. 3.11 The periodic signal $\tilde{x}[n]$ for Example 3.9

$$x[n]$$

$$\cdots a \quad b \quad c \quad d \quad a \quad b \quad c \quad d \quad a \cdots$$

$$-4 \quad -3 \quad -2 \quad -1 \quad 0 \quad 1 \quad 2 \quad 3 \quad 4 \qquad n$$

Calculation of the periodic signal $\tilde{x}[n_0 - n]$ can be achieved via the following steps.

(1) We first find one period of $\tilde{x}_1[n] = \tilde{x}[-n]$ using rotate inside operation.
(2) Then one period of $\tilde{x}_1[n]$ is rotated to the right if $n_0 > 0$ to the left if $n_0 < 0$ by $|n_0|$ units and one period of $\tilde{x}_1[n_0 - n]$ is obtained.

Example 3.9 The periodic signal $\tilde{x}_1[n]$ is shown in Fig. 3.11. Find $\tilde{x}_1[2 - n]$.

Solution 3.9 From Fig. 3.11 one period of $\tilde{x}[n]$ can be found as

$$x[n] = [a \quad b \quad c \quad d]. \tag{3.20}$$

When (3.20) is rotated inside, we obtain

$$RR(x[n]) = [a \quad d \quad c \quad b] \tag{3.21}$$

which is nothing but one period of $\tilde{x}[-n]$, i.e., $\tilde{x}[-n]_{op} = [a \quad d \quad c \quad b]$, 'op' means one period. To find one period of $\tilde{x}[2 - n]$ one period of $\tilde{x}[-n]$ is rotated to the right by 2 units yielding

$$RR\left(\tilde{x}[-n]_{op}, 2\right) = [a \quad b \quad c \quad d]. \tag{3.22}$$

Using (3.22) the periodic signal $\tilde{x}[2 - n]$ can be written as

$$\tilde{x}[2 - n] = [\cdots c \quad b \quad a \quad d \quad \underbrace{c}_{n=0} \quad b \quad a \quad d \quad c \quad b \quad a \quad d \cdots].$$

Example 3.10 The periodic signal $\tilde{x}[n]$ is shown in Fig. 3.11. Find $\tilde{x}[-2 - n]$. Fig. 3.12

Solution 3.10 One period of $\tilde{x}[n]$ equals to

$$x[n] = [a \quad b \quad c \quad d]. \tag{3.23}$$

Fig. 3.12 The periodic signal $\tilde{x}[n]$ for Example 3.10

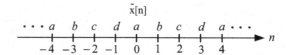

$$x[n]$$

$$\cdots a \quad b \quad c \quad d \quad a \quad b \quad c \quad d \quad a \cdots$$

$$-4 \quad -3 \quad -2 \quad -1 \quad 0 \quad 1 \quad 2 \quad 3 \quad 4 \qquad n$$

When (3.23) is rotated inside, we obtain

$$RR(x[n]) = [a \quad d \quad c \quad b]$$

which is nothing but one period of $\tilde{x}[-n]$, i.e., $\tilde{x}[-n]_{op} = [a \quad d \quad c \quad b]$. To find one period of $\tilde{x}[-2 - n]$, one period of $\tilde{x} - n]$ is rotated to the left by 2 units yielding

$$RR\left(\tilde{x}[-n]_{op}, 2\right) = [c \quad b \quad a \quad d]. \tag{3.24}$$

Using (3.24) the periodic signal $\tilde{x}[-2 - n]$ can be written as

$$\tilde{x}[-2 - n] = [\cdots c \quad b \quad a \quad d \quad \underset{n=0}{\underbrace{c}} \quad b \quad a \quad d \quad c \quad b \quad a \quad d \cdots].$$

Exercise: For the previous exercise find $\tilde{x}[-4 - n]$ and $\tilde{x}[4 - n]$.

3.1.3   Some Well Known Digital Signals

In this subsection, we will review some well-known digital signals.
Unit Step:
The unit step signal is defined as

$$u[n] = \begin{cases} 1 & \text{if } n \geq 0 \\ 0 & \text{otherwise} \end{cases} \tag{3.25}$$

whose graph is shown in Fig. 3.13.
Unit Impulse:
The unit impulse signal is defined as

$$\delta[n] = \begin{cases} 1 & \text{if } n = 0 \\ 0 & \text{otherwise} \end{cases} \tag{3.26}$$

whose graph is shown in Fig. 3.14.

Fig. 3.13 Unit step function, i.e., signal

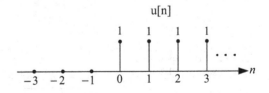

Fig. 3.14 Unit impulse
function, i.e., signal

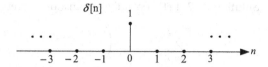

The relation between $u[n]$ and $\delta[n]$ can be written as

$$\delta[n] = u[n] - u[n-1] \tag{3.27}$$

or as

$$u[n] = \sum_{k=0}^{\infty} \delta[n-k] \tag{3.28}$$

which is equal to

$$u[n] = \sum_{k=-\infty}^{n} \delta[k]. \tag{3.29}$$

Exponential Digital Signal

The exponential digital signal is defined as

$$x[n] = e^{jw_0 n} \tag{3.30}$$

which can also be written in the form

$$x[n] = \cos(w_0 n) + j\sin(w_0 n). \tag{3.31}$$

Example 3.11 Simplify $e^{-jk2\pi}$.

Solution 3.11 Using (3.31), we have

$$e^{-jk2\pi} = \cos(-k2\pi) + j\sin(-k2\pi) = \underbrace{\cos(k2\pi)}_{=1} + j\underbrace{\sin(-k2\pi)}_{=0}$$

As a special case for $k = 1$, we have $e^{\pm j2\pi} = 1$.

Example 3.12 Verify the following equality

$$\sum_{k=0}^{N-1} e^{-j\frac{2\pi}{N}km} = \begin{cases} N & \text{if } m = 0 \\ 0 & \text{otherwise.} \end{cases} \tag{3.32}$$

Solution 3.12 Let's open the summation expression in (3.32) as follows

$$\sum_{k=0}^{N-1} e^{-j\frac{2\pi}{N}km} = \left(1 + e^{-j\frac{2\pi}{N}m} + e^{-j\frac{2\pi}{N}2m} + \cdots + e^{-j\frac{2\pi}{N}(N-1)m}\right). \tag{3.33}$$

The right hand side of (3.33) can be simplified using the property

$$1 + x + x^2 + x^3 + \cdots + x^{N-1} = \frac{1-x^N}{1-x} \tag{3.34}$$

as in

$$\left(1 + e^{-j\frac{2\pi}{N}m} + e^{-j\frac{2\pi}{N}2m} + \cdots + e^{-j\frac{2\pi}{N}(N-1)m}\right) = \frac{1 - e^{-j\frac{2\pi}{N}mN}}{1 - e^{-j\frac{2\pi}{N}m}} = \frac{1 - e^{-j2\pi m}}{1 - e^{-j\frac{2\pi}{N}m}}. \tag{3.35}$$

And for $m \neq 0$ using the result in, (3.35), we obtain

$$\frac{1 - e^{-j2\pi m}}{1 - e^{-j\frac{2\pi}{N}m}} = \frac{1-1}{1 - e^{-j\frac{2\pi}{N}m}} \to 0. \tag{3.36}$$

Hence we have

$$\sum_{k=0}^{N-1} e^{-j\frac{2\pi}{N}km} = 0, \quad m \neq 0. \tag{3.37}$$

And for $m = 0$ using the result in (3.35), we obtain

$$\sum_{k=0}^{N-1} e^{-j\frac{2\pi}{N}km} = \sum_{k=0}^{N-1} 1 \to N. \tag{3.38}$$

Combining (3.37) and (3.38), we obtain

$$\sum_{k=0}^{N-1} e^{\pm j\frac{2\pi}{N}km} = \begin{cases} N & if \ m = 0 \\ 0 & otherwise. \end{cases} \tag{3.39}$$

3.2 Review of Signal Types

Basically we can divide signals into two categories as, continuous and digital signals. And in both classes, we can have periodic and non-periodic (aperiodic) signals, and Fourier transform and representation methods are defined for these classes of signals. In Fig. 3.15; the relation between signals and their transform or representation types are summarized.

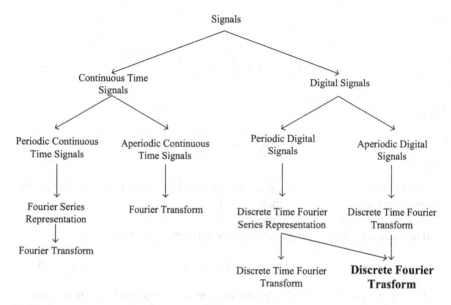

Fig. 3.15   Signals types, their transformations and representations

Let's briefly review the signal types, their transformations and representations.
Non-periodic Continuous Time Signals
If $x_c(t)$ is a non-periodic continuous time signal, then its Fourier is defined as

$$X_c(w) = \int_{-\infty}^{\infty} x_c(t)e^{-jwt}dt \tag{3.40}$$

and its inverse Fourier transform is given as

$$x_c(t) = \frac{1}{2\pi} \int_{-\infty}^{\infty} X_c(w)e^{jwt}dw \tag{3.41}$$

where $w = 2\pi f$ is the angular frequency. The Fourier transform and inverse Fourier transform pairs show small differences in their coefficients in literature. In general, Fourier transform and inverse Fourier transform can be defined as

$$X_c(w) = K_1 \int_{-\infty}^{\infty} x_c(t)e^{-jwt}dt \tag{3.42}$$

and

$$x_c(t) = K_2 \int\limits_{-\infty}^{\infty} X_c(w)e^{jwt}dw \qquad (3.43)$$

where

$$K_1 \times K_2 = \frac{1}{2\pi}. \qquad (3.44)$$

Thus if $K_1 = 1/\sqrt{2\pi}$, then K_2 should be $1/\sqrt{2\pi}$ so that $K_1 \times K_2 = 1/2\pi$. As another example if $K_1 = 1/2\pi$ then $K_2 = 1$.

Periodic Continuous Time Signals

If $\tilde{x}_c(t)$ is a periodic signal with fundamental period T, then

$$\tilde{x}_c(t) = \tilde{x}_c(t + mT). \qquad (3.45)$$

And for the periodic signal $\tilde{x}_c(t)$ the Fourier series representation is defined as

$$\tilde{x}_c(t) = \frac{1}{T} \sum_{k=-\infty}^{\infty} \tilde{x}[k]e^{jk\frac{2\pi}{T}t} \qquad (3.46)$$

where the Fourier series coefficients $\tilde{x}[k]$ are computed by using

$$\tilde{x}_c[k] = \int\limits_{T} \tilde{x}_c(t)e^{-jk\frac{2\pi}{T}t}dt. \qquad (3.47)$$

If we define $2\pi/T$ by w_0, i.e., $w_0 = 2\pi/T$, then the above equations can also be written as

$$\tilde{x}_c(t) = \frac{1}{T} \sum_{k=-\infty}^{\infty} \tilde{x}_c[k]e^{jkw_0t} \qquad (3.48)$$

and

$$\tilde{x}_c[k] = \int\limits_{T} \tilde{x}_c(t)e^{-jkw_0t}dt \qquad (3.49)$$

In general, the Fourier series representation of $\tilde{x}_c(t)$ and its Fourier series coefficients are given as

$$\tilde{x}_c(t) = K_1 \sum_{k=-\infty}^{\infty} \tilde{x}_c[k] e^{jkw_0 t} \qquad (3.50)$$

and

$$\tilde{x}_c[k] = K_2 \int_T \tilde{x}_c(t) e^{-jkw_0 t} dt \qquad (3.51)$$

where the coefficients satisfy $K_1 \times K_2 = 1/T$. Hence, if $K_1 = 1/\sqrt{T}$ then $K_2 = 1/\sqrt{T}$ and Fourier series representation and Fourier coefficients expressions becomes as

$$\tilde{x}_c(t) = \frac{1}{\sqrt{T}} \sum_{k=-\infty}^{\infty} \tilde{x}_c[k] e^{jkw_0 t} \qquad (3.52)$$

and

$$\tilde{x}_c[k] = \frac{1}{\sqrt{T}} \int_T \tilde{x}_c(t) e^{-jkw_0 t} dt. \qquad (3.53)$$

Now let's assume that one period of $\tilde{x}_c(t)$ is $x_c(t)$, i.e., $x_c(t)$ is an aperiodic signal. Then the Fourier series coefficients of $\tilde{x}_c(t)$ is computed as

$$\tilde{x}_c[k] = \int_T \tilde{x}_c(t) e^{-jkw_0 t} dt \rightarrow \tilde{x}_c[k] = \int_{\infty}^{\infty} x_c(t) e^{-jkw_0 t} dt. \qquad (3.54)$$

And the Fourier transform of $x_c(t)$ is

$$X_c(w) = \int_{-\infty}^{\infty} x_c(t) e^{-jwt} dt \qquad (3.55)$$

When (3.54) and (3.55) are compared to each other as in

$$\tilde{x}_c[k] = \int_{\infty}^{\infty} x_c(t) e^{-jkw_0 t} dt \leftrightarrow X_c(w) = \int_{-\infty}^{\infty} x_c(t) e^{-jwt} dt \qquad (3.56)$$

we see that

$$\tilde{x}_c[k] = X_c(w)|_{w=kw_0} \qquad (3.57)$$

where

$$w_0 = \frac{2\pi}{T}. \qquad (3.58)$$

And the relation between $\tilde{x}_c(t)$ and $x_c(t)$ can be written as

$$\tilde{x}_c(t) = \sum_{k=-\infty}^{\infty} x_c(t - kT). \qquad (3.59)$$

The Fourier transform of the periodic continuous time signal is defined as

$$\tilde{x}_c(w) = \frac{2\pi}{T} \sum_{k=-\infty}^{\infty} \tilde{x}_c[k]\delta(w - kw_0), \quad w_0 = 2\pi/T. \qquad (3.60)$$

Aperiodic Digital Signals
The discrete time Fourier transform for the aperiodic digital signal $x[n]$ is defined as

$$X_n(w) = \sum_{n=-\infty}^{\infty} x[n]e^{-jwn} \qquad (3.61)$$

where $w = 2\pi f$ is the angular frequency, and the inverse Fourier transform is defined as

$$x[n] = \frac{1}{2\pi} \int_{2\pi} X_n(w)e^{jwn}dw. \qquad (3.62)$$

The Fourier transform function of $x[n]$, i.e., $X_n(w)$ is a continuous function of w and it is also a periodic function with period 2π, i.e.,

$$X_n(w) = X_n(w + k2\pi). \qquad (3.63)$$

Periodic Digital Signals
If the digital signal $\tilde{x}[n]$ is a periodic signal, then $\tilde{x}[n] = \tilde{x}[n + lN]$ $l, N \in Z$ and N is called fundamental period of $\tilde{x}[n]$.

For the digital periodic signal $\tilde{x}[n]$, the Fourier series representation is defined as

$$\tilde{x}[n] = \frac{1}{N} \sum_{k,N} \tilde{x}_n[k] e^{jk\frac{2\pi}{N}n} \tag{3.64}$$

where the Fourier series coefficients are computed using

$$\tilde{x}_n[k] = \sum_{n,N} \tilde{x}[n] e^{-jk\frac{2\pi}{N}n}. \tag{3.65}$$

Note: $\sum_{n,N}(\cdot)$ means summation is taken over any interval of length N, i.e., summation is taken over one period length.

In general, the Fourier series representation and calculation of Fourier series coefficient of periodic signals are done via

$$\tilde{x}[n] = K_1 \sum_{k,N} \tilde{x}_n[k] e^{jk\frac{2\pi}{N}n} \tag{3.66}$$

and

$$\tilde{x}_n[k] = K_2 \sum_{n,N} \tilde{x}[n] e^{-jk\frac{2\pi}{N}n} \tag{3.67}$$

such that

$$K_1 \times K_2 = \frac{1}{N}. \tag{3.68}$$

The Fourier transform of the periodic digital signal $\tilde{x}[n]$ is

$$\tilde{x}(w) = \frac{2\pi}{N} \sum_{k=-\infty}^{\infty} \tilde{x}_n[k] \delta(w - kw_0), \quad w_0 = \frac{2\pi}{N}. \tag{3.69}$$

Example 3.13 If the Fourier series representation of digital periodic signal $\tilde{x}[n]$ is

$$\tilde{x}[n] = \frac{1}{N} \sum_{k,N} \tilde{x}_n[k] e^{jk\frac{2\pi}{N}n} \tag{3.70}$$

then verify that the Fourier series coefficients as obtained using

$$\tilde{x}_n[k] = \sum_{n,N} \tilde{x}[n] e^{-jk\frac{2\pi}{N}n}. \tag{3.71}$$

Solution 3.13 If the Fourier series coefficients are obtained using

$$\tilde{x}_n[k] = \sum_{n,N} \tilde{x}[n] e^{-jk\frac{2\pi}{N}n} \tag{3.72}$$

then when (3.72) is substituted into

$$\tilde{x}[n] = \frac{1}{N} \sum_{k,N} \tilde{x}_n[k] e^{jk\frac{2\pi}{N}n} \tag{3.73}$$

we should get $\tilde{x}[n]$ on the right hand side of (3.73). That is

$$
\begin{aligned}
\tilde{x}[n] &= \frac{1}{N} \sum_{k,N} \underbrace{\sum_{r,N} \tilde{x}[r] e^{-jk\frac{2\pi}{N}r}}_{\tilde{x}_n[k]} e^{jk\frac{2\pi}{N}n} \\
&= \frac{1}{N} \sum_{k=0}^{N-1} \sum_{r=0}^{N-1} \tilde{x}[r] e^{-jk\frac{2\pi}{N}r} e^{jk\frac{2\pi}{N}n} \\
&= \frac{1}{N} \sum_{k=0}^{N-1} \sum_{r=0}^{N-1} \tilde{x}[r] e^{-jk\frac{2\pi}{N}(r-n)} \\
&= \frac{1}{N} \sum_{r=0}^{N-1} \tilde{x}[r] \underbrace{\sum_{k=0}^{N-1} e^{-jk\frac{2\pi}{N}(r-n)}}_{=\begin{cases} N & \text{if } r = n \\ 0 & \text{otherwise} \end{cases}} \\
&= \frac{1}{N} N\tilde{x}[n] \\
&= \tilde{x}[n]
\end{aligned}
\tag{3.74}
$$

Convolution of Aperiodic Digital Signals

For aperiodic digital signals $x[n], y[n]$, the convolution operation is defined as

$$x[n] * y[n] = \sum_{k=-\infty}^{\infty} x[k] y[n-k] \tag{3.75}$$

or

$$x[n] * y[n] = \sum_{k=-\infty}^{\infty} x[n-k] y[k]. \tag{3.76}$$

3.3 Convolution of Periodic Digital Signals

Let $\tilde{x}_n[n]$ and $\tilde{x}_2[n]$ be digital periodic signals with common period N, i.e., $\tilde{x}_1[n] = \tilde{x}_1[n+N]$ and $\tilde{x}_2[n] = \tilde{x}_2[n+N]$.

The period convolution of $\tilde{x}_1[n]$ and $\tilde{x}_2[n]$ is defined as

$$\tilde{x}_3[n] = \sum_{m=0}^{N-1} \tilde{x}_1[m]\tilde{x}_2[n-m]. \tag{3.77}$$

The digital sequence $\tilde{x}_3[n]$ is also periodic with period N. How to calculate periodic convolution? This is explained as follows.

(1) Since $\tilde{x}_3[n]$ is periodic with the same period N, we can focus on the calculation of one period of $\tilde{x}_3[n]$ starting from 0, i.e., consider $0 \le n \le N-1$.
(2) When the summation in (3.77) is expanded, we get

$$\tilde{x}_3[n] = \tilde{x}_1[0]\tilde{x}_2[n] + \tilde{x}_1[1]\tilde{x}_2[n-1] + \cdots + \tilde{x}_1[N-1]\tilde{x}_2[n-(N-1)] \tag{3.78}$$

where we can use only one period of $\tilde{x}_2[n]$, $\tilde{x}_2[n-1]$, and $\tilde{x}_2[N-1]$, $0 \le n \le N-1$.

Example 3.14 The periodic signals $\tilde{x}_1[n]$ and $\tilde{x}_2[n]$ with period $N = 4$ are shown in Fig. 3.16. Calculate their 4-point periodic convolution.

Solution 3.14 The periodic convolution for the given signals is calculated using

$$\tilde{x}_3[n] = \sum_{m=0}^{N-1} \tilde{x}_1[m]\tilde{x}_2[n-m]. \tag{3.79}$$

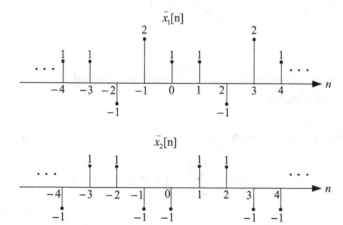

Fig. 3.16 The periodic signals $\tilde{x}_1[n]$ and $\tilde{x}_2[n]$ for Example 3.14

When the summation in (3.79) is expanded for $N = 4$, we get

$$\tilde{x}_3[n] = \tilde{x}_1[0]\tilde{x}_2[n] + \tilde{x}_1[1]\tilde{x}_2[n-1] + \tilde{x}_1[2]\tilde{x}_2[n-2] + \tilde{x}_1[3]\tilde{x}_2[n-3]. \qquad (3.80)$$

One period of $\tilde{x}_2[n], \tilde{x}_2[n-1], \tilde{x}_2[n-2]$, and $\tilde{x}_2[n-3]$ for $0 \leq n \leq 3$ can be calculated using rotate right operation yielding

$$\begin{aligned}
\tilde{x}_{2op}[n] &= \begin{bmatrix} -1 & 1 & 1 & -1 \end{bmatrix} \\
\tilde{x}_{2op}[n-1] &= \begin{bmatrix} -1 & -1 & 1 & 1 \end{bmatrix} \\
\tilde{x}_{2op}[n-2] &= \begin{bmatrix} 1 & -1 & -1 & 1 \end{bmatrix} \\
\tilde{x}_{2op}[n-3] &= \begin{bmatrix} 1 & 1 & -1 & -1 \end{bmatrix}.
\end{aligned} \qquad (3.81)$$

Substituting (3.81) into (3.80), one period of $\tilde{x}_3[n]$ is calculated as

$$\begin{aligned}
\tilde{x}_{3op}[n] = {}& \tilde{x}_1[0]\tilde{x}_{2op}[n] + \tilde{x}_1[1]\tilde{x}_{2op}[n-1] + \tilde{x}_1[2]\tilde{x}_{2op}[n-2] \\
& + \tilde{x}_1[3]\tilde{x}_{2op}[n-3]
\end{aligned}$$

yielding

$$\begin{aligned}
\tilde{x}_{3op}[n] = {}& 1 \times \begin{bmatrix} -1 & 1 & 1 & -1 \end{bmatrix} + 1 \times \begin{bmatrix} -1 & -1 & 1 & 1 \end{bmatrix} - 1 \times \begin{bmatrix} 1 & -1 & -1 & 1 \end{bmatrix} \\
& + 2 \times \begin{bmatrix} 1 & 1 & -1 & -1 \end{bmatrix}
\end{aligned}$$

which can be simplified as

$$\tilde{x}_{3op}[n] = \begin{bmatrix} -1 & 3 & 1 & -3 \end{bmatrix}. \qquad (3.82)$$

Using (3.82), the periodic convolution result can be written as

$$\tilde{x}_3[n] = \begin{bmatrix} \cdots -1 & 3 & 1 -3 & \underbrace{-1}_{n=0} & 3 & 1 -3 & -1 & 3 & 1 & -3 & \cdots \end{bmatrix}.$$

3.3.1 Alternative Method to Compute the Periodic Convolution

The periodic convolution expression

$$\tilde{x}_3[n] = \sum_{m=0}^{N-1} \tilde{x}_1[m]\tilde{x}_2[n-m] \qquad (3.83)$$

can be computed for $n = 0, 1, \ldots, N - 1$ as

$$n = 0, \quad \tilde{x}_3[0] = \sum_{m=0}^{N-1} \tilde{x}_1[m]\tilde{x}_2[-m]$$

$$n = 1, \quad \tilde{x}_3[1] = \sum_{m=0}^{N-1} \tilde{x}_1[m]\tilde{x}_2[1 - m]$$

$$n = 2, \quad \tilde{x}_3[2] = \sum_{m=0}^{N-1} \tilde{x}_1[m]\tilde{x}_2[2 - m]$$

$$\vdots$$

$$n = N - 1, \quad \tilde{x}_3[N - 1] = \sum_{m=0}^{N-1} \tilde{x}_1[m]\tilde{x}_2[(N - 1) - m].$$

Now let's consider

$$\tilde{x}_3[0] = \sum_{m=0}^{N-1} \tilde{x}_1[m]\tilde{x}_2[-m] \tag{3.84}$$

when expanded for $N = 3$, we get

$$\tilde{x}_3[0] = \tilde{x}_1[0]\tilde{x}_2[0] + \tilde{x}_1[1]\tilde{x}_2[-1] + \tilde{x}_1[2]\tilde{x}_2[-2] + \tilde{x}_1[3]\tilde{x}_2[-3] \tag{3.85}$$

Since $\tilde{x}_3[n] = \tilde{x}_3[n + 4]$, we have

$$\tilde{x}_2[-1] = \tilde{x}_2[3], \quad \tilde{x}_2[-2] = \tilde{x}_2[2], \quad \tilde{x}_2[-3] = \tilde{x}_2[1]. \tag{3.86}$$

Using (3.86) in (3.85), we obtain

$$\tilde{x}_3[0] = \tilde{x}_1[0]\tilde{x}_2[0] + \tilde{x}_1[1]\tilde{x}_2[3] + \tilde{x}_1[2]\tilde{x}_2[2] + \tilde{x}_1[3]\tilde{x}_2[1] \tag{3.87}$$

which can be written as the dot product of the vectors

$$[\tilde{x}_1[0] \quad \tilde{x}_1[1] \quad \tilde{x}_1[2] \quad \tilde{x}_1[3]] \text{ and } [\tilde{x}_2[0] \quad \tilde{x}_2[0] \quad \tilde{x}_2[2] \quad \tilde{x}_2[1]]$$

where it is clear that the vector $[\tilde{x}_2[0] \quad \tilde{x}_2[3] \quad \tilde{x}_2[3] \quad \tilde{x}_2[1]]$ can be obtained from one period of $\tilde{x}_2[n]$ via rotate inside operation.

Hence we can write

$$\tilde{x}_3[0] = \sum_{m=0}^{N-1} \tilde{x}_1[m]\tilde{x}_2[-m] \rightarrow \tilde{x}_3[0] = \tilde{x}_{1op}[m] * \tilde{x}_{2op}[-m] \tag{3.88}$$

$$\tilde{x}_3[1] = \sum_{m=0}^{N-1} \tilde{x}_1[m]\tilde{x}_2[1 - m] \rightarrow \tilde{x}_3[1] = \tilde{x}_{1op}[m] * \tilde{x}_{2op}[1 - m]. \tag{3.89}$$

Equation (3.89) can be written as

$$\tilde{x}_3[1] = \sum_{m=0}^{N-1} \tilde{x}_1[m]\tilde{x}_2[1-m] \rightarrow \tilde{x}_3[1] = \tilde{x}_{1op}[m] * RR\big(\tilde{x}_{2op}[-m]\big) \tag{3.90}$$

and in a similar manner

$$\tilde{x}_3[2] = \sum_{m=0}^{N-1} \tilde{x}_1[m]\tilde{x}_2[2-m] \rightarrow \tilde{x}_3[2] = \tilde{x}_{1op}[m] * RR\big(\tilde{x}_{2op}[1-m]\big) \tag{3.91}$$

$$\tilde{x}_3[3] = \sum_{m=0}^{N-1} \tilde{x}_1[m]\tilde{x}_2[3-m] \rightarrow \tilde{x}_3[3] = \tilde{x}_{1op}[m] * RR\big(\tilde{x}_{2op}[2-m]\big) \tag{3.92}$$

$$\vdots$$

$$\tilde{x}_3[N-1] = \sum_{m=0}^{N-1} \tilde{x}_1[m]\tilde{x}_2[N-1-m] \rightarrow$$

$$\tilde{x}_3[N-1] = \tilde{x}_{1op}[m] * RR\big(\tilde{x}_{2op}[N-2-m]\big) \tag{3.93}$$

Example 3.15 The periodic signals $\tilde{x}_1[n]$ and $\tilde{x}_2[n]$ with period $= 4$ are shown in Fig. 3.17. Calculate their 4-point periodic convolution using alternative periodic convolution method.

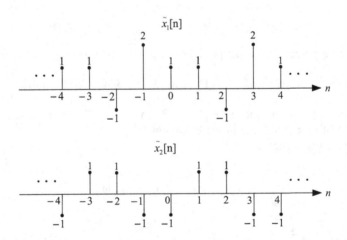

Fig. 3.17 The periodic signals $\tilde{x}_1[n]$ and $\tilde{x}_2[n]$ for Example 3.15

Solution 3.15 When the periodic convolution expression

$$\tilde{x}_3[n] = \sum_{m=0}^{N-1} \tilde{x}_1[m]\tilde{x}_2[n-m]$$

is calculated for $n = 0, 1, \ldots, N-1$, we get

$$\tilde{x}_3[0] = \tilde{x}_{1op}[m] * \tilde{x}_{2op}[-m]$$
$$\tilde{x}_3[1] = \tilde{x}_{1op}[m] * \tilde{x}_{2op}[1-m]$$
$$\tilde{x}_3[2] = \tilde{x}_{1op}[m] * RR(\tilde{x}_{2op}[1-m])$$
$$\tilde{x}_3[3] = \tilde{x}_{1op}[m] * RR(\tilde{x}_{2op}[2-m]).$$

(3.94)

One period of $\tilde{x}_2[n]$ for $0 \le n \le 3$ is

$$\tilde{x}_{2op}[m] = \begin{bmatrix} -1 & 1 & 1 & -1 \end{bmatrix}.$$

Then

$$\tilde{x}_{2op}[-m] = \begin{bmatrix} -1 & -1 & 1 & 1 \end{bmatrix}$$
$$\tilde{x}_{2op}[1-m] = RR(\tilde{x}_{2op}[-m]) \to RR(\tilde{x}_{2op}[-m]) = \begin{bmatrix} 1 & -1 & -1 & 1 \end{bmatrix}$$
$$\tilde{x}_{2op}[2-m] = RR(\tilde{x}_{2op}[1-m]) \to RR(\tilde{x}_{2op}[-m]) = \begin{bmatrix} 1 & 1 & -1 & -1 \end{bmatrix}$$
$$\tilde{x}_{2op}[3-m] = RR(\tilde{x}_{2op}[2-m]) \to RR(\tilde{x}_{2op}[-m]) = \begin{bmatrix} -1 & 1 & 1 & -1 \end{bmatrix}$$

(3.95)

and

$$\tilde{x}_{1op}[m] = \begin{bmatrix} 1 & 1 & -1 & 2 \end{bmatrix}.$$

(3.96)

Using (3.95) and (3.96) in (3.94), we can calculate the periodic convolution values as

$$\tilde{x}_3[0] = \begin{bmatrix} 1 & 1 & -1 & 2 \end{bmatrix} * \begin{bmatrix} -1 & -1 & 1 & 1 \end{bmatrix} \to$$
$$\tilde{x}_3[0] = 1(-1) + 1(-1) + (-1)1 + 2 \times 1 \to$$
$$\tilde{x}_3[3] = -1$$

$$\tilde{x}_3[1] = \begin{bmatrix} 1 & 1 & -1 & 2 \end{bmatrix} * \begin{bmatrix} 1 & -1 & -1 & 1 \end{bmatrix} \to \tilde{x}_3[1] = 3$$
$$\tilde{x}_3[2] = \begin{bmatrix} 1 & 1 & -1 & 2 \end{bmatrix} * \begin{bmatrix} 1 & 1 & -1 & -1 \end{bmatrix} \to \tilde{x}_3[2] = 1$$
$$\tilde{x}_3[3] = \begin{bmatrix} 1 & 1 & -1 & 2 \end{bmatrix} * \begin{bmatrix} -1 & 1 & 1 & -1 \end{bmatrix} \to \tilde{x}_3[3] = -3$$

Hence,

$$\tilde{x}_{3op}[n] = [-1 \quad 3 \quad 1 \quad -3].$$

Then the periodic convolution result becomes as

$$\tilde{x}_3[n] = [\cdots \quad -1 \quad 3 \quad 1 \quad -3 \quad \underbrace{-1}_{n=0} \quad 3 \quad 1 \quad -3 \quad 1 \quad 3 \quad 1 \quad -3 \quad \cdots].$$

3.4 Sampling of Fourier Transform

The Fourier transform $X_n(w)$ of a non-periodic digital signal $x[n]$ is a continuous function of w and it is periodic with period 2π, i.e., $X_n(w) = X_n(w + 2\pi)$.

Example 3.16 The Fourier transform of the signal $x[n] = \frac{1}{2}\delta[n+1] + \frac{1}{2}\delta[n-1]$ is calculated as

$$
\begin{aligned}
X_n(w) &= \sum_{n=-\infty}^{\infty} x[n]e^{-jwn} \\
&= \sum_{n=-\infty}^{\infty} \left(\frac{1}{2}\delta[n+1] + \frac{1}{2}\delta[n-1] \right) e^{-jwn} \\
&= \frac{1}{2} \left(e^{jw} + e^{-jw} \right) \\
&= \cos(w).
\end{aligned}
\tag{3.97}
$$

The aperiodic digital signal $x[n]$ and its Fourier transform is shown in Fig. 3.18. Let's generate the periodic signal $\tilde{x}[n]$ with period N from $x[n]$ via

$$\tilde{x}[n] = \sum_{l=-\infty}^{\infty} x[n - lN]. \tag{3.98}$$

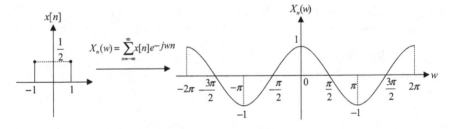

Fig. 3.18 The aperiodic digital signal $x[n]$ in Example 3.16 and its Fourier transform

The Fourier series coefficients of the periodic signal $\tilde{x}[n]$ in (3.98) are obtained from the Fourier transform of $x[n]$, i.e., $X_n(w)$, via sampling operation in frequency domain as in

$$\tilde{X}[k] = X_n(w)|_{w=kw_s} \tag{3.99}$$

where $w_s = \frac{2\pi}{N}$ is the sampling period in radian unit.

Example 3.17 $\tilde{x}[n]$ is a periodic signal with period $N = 4$, and we have $x[n] = \frac{1}{2}\delta[n+1] + \frac{1}{2}\delta[n-1]$ for one period of this signal. In addition, the periodic signal can be obtained from its one period via

$$\tilde{x}[n] = \sum_{l=-\infty}^{\infty} x[n - lN]. \tag{3.100}$$

Find the Fourier series coefficients of $\tilde{x}[n]$ using $X_n(w)$ the Fourier transform of $x[n]$.

Solution 3.17 In Example 3.17, we found the Fourier transform of $x[n] = \frac{1}{2}\delta[n+1] + \frac{1}{2}\delta[n-1]$ as

$$X_n(w) = \cos(w).$$

The Fourier series coefficients, i.e., $\tilde{X}[k]$, of $\tilde{x}[n]$ can be obtained via sampling operation in frequency using

$$\tilde{X}[k] = X_n(w)|_{w=kw_s} \tag{3.101}$$

where $w_s = \frac{2\pi}{N} \rightarrow w_s = \frac{2\pi}{4} \rightarrow w_s = \frac{\pi}{2}$. Hence (3.101) yields

$$\tilde{X}[k] = X_n(w)|_{w=kw_s} \rightarrow \tilde{X}[k] = \cos(w)|_{w=\frac{k\pi}{2}} \rightarrow \tilde{X}[k] = \cos\left(\frac{k\pi}{2}\right). \tag{3.102}$$

The graphical illustration of the sampling operation in frequency domain is explained in Fig. 3.19.

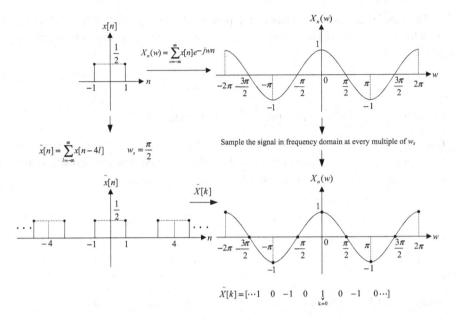

Fig. 3.19 Fourier series coefficients are obtained from Fourier transform via sampling operation

3.5   Discrete Fourier Transform

The Fourier series coefficients, i.e., $\tilde{X}[k]$, is a periodic function which can have complex or real values. The Fourier series coefficients $\tilde{X}[k]$ satisfy $\tilde{X}[k] = \tilde{X}[k+N]$ where N is the period of the digital signal $\tilde{x}[n]$.

The periodic signal $\tilde{x}[n]$ with period N has the Fourier series coefficients

$$\tilde{X}[k] = \sum_{n=0}^{N-1} \tilde{x}[n]e^{-j\frac{2\pi}{N}kn} \qquad (3.103)$$

and for $0 \le n < N$, $\tilde{x}[n] = x[n]$ where $x[n]$ is one period of $\tilde{x}[n]$. Then (3.103) can be written as

$$\tilde{X}[k] = \sum_{n=0}^{N-1} x[n]e^{-j\frac{2\pi}{N}kn} \qquad (3.104)$$

which is also a periodic signal with the same period as the time domain signal $\tilde{x}[n]$. Let's consider one period of $\tilde{X}[k]$

$$X[k] = \begin{cases} \tilde{X}[k] & if\ 0 \le k < N \\ 0 & otherwise \end{cases} \tag{3.105}$$

which is called the discrete Fourier transform of $x[n]$. Thus, $N - point$ discrete Fourier transform of $x[n]$ is defined as

$$X[k] = \sum_{n=0}^{N-1} x[n] e^{-j\frac{2\pi}{N}kn}, \quad 0 \le k < N. \tag{3.106}$$

Similarly, $N - point$ inverse Fourier transform is defined as

$$x[n] = \frac{1}{N} \sum_{k=0}^{N-1} X[k] e^{j\frac{2\pi}{N}kn}, \quad 0 \le n < N.$$

A more general definition for N-point DFT is

$$X[k] = \sum_{n,N} x[n] e^{-j\frac{2\pi}{N}kn}, \quad k, N. \tag{3.107}$$

and for the N-point inverse DFT, a more general definition is

$$x[n] = \frac{1}{N} \sum_{k,N} X[k] e^{j\frac{2\pi}{N}kn}, \quad n, N. \tag{3.108}$$

In addition, Fourier series coefficients of a periodic signal can be obtained from the Fourier transform of its one period using

$$\tilde{X}[k] = X_n(w)|_{w=kw_s} \quad w_s = \frac{2\pi}{N}. \tag{3.109}$$

And using the definition

$$X[k] = \begin{cases} \tilde{X}[k] & if\ 0 \le k < N \\ 0 & otherwise \end{cases} \tag{3.110}$$

we can write

$$X[k] = X_n(w)|_{w=kw_s}, \quad w_s = \frac{2\pi}{N}, \ 0 \le k < N \tag{3.111}$$

which means that the discrete Fourier transform of $x[n]$ is nothing but a mathematical sequence obtained from one period of $X_n(w)$ via sampling operation in frequency domain, and the sampling period is chosen as $w_s = \frac{2\pi}{N}$.

Example 3.18 Find the discrete Fourier transform of

$$x[n] = [1 \quad 1 \quad -1 \quad 2].$$

Solution 3.18 For the given signal if the DFT formula

$$X[k] = \sum_{n=0}^{4-1} x[n] e^{-j\frac{2\pi}{4}kn}, \quad 0 \leq k < 4 \tag{3.112}$$

is expanded, the coefficients are found as

$$X[k] = \underbrace{x[0]}_{1} \times e^{0} + \underbrace{x[1]}_{1} \times e^{-j\frac{2\pi}{4}k} + \underbrace{x[2]}_{-1} \times e^{-j\frac{2\pi}{4}k2} + \underbrace{x[3]}_{2} \times e^{-j\frac{2\pi}{4}k3}. \tag{3.113}$$

When (3.113) is simplified, we obtain

$$X[k] = 1 + e^{-j\frac{\pi}{2}k} - 1e^{-j\pi k} + 2e^{-j\frac{3\pi}{2}k} \tag{3.114}$$

Evaluating (3.114), i.e., $X[k]$, for $k = 0, 1, 2, 3$, we get

$$X[0] = 3 \quad X[1] = 2 + j \quad X[2] = -3 \quad X[3] = 2 - j$$

which can be written in short as

$$X[k] = [3 \quad 2 + j \quad -3 \quad 2 - j].$$

Example 3.19 Find the aperiodic digital signal whose DFT coefficients are given as

$$X[k] = [3 \quad 2 + j \quad -3 \quad 2 - j].$$

Solution 3.19 Using $X[k]$ in inverse DFT formula

$$x[n] = \frac{1}{4} \sum_{k=0}^{4-1} X[k] e^{j\frac{2\pi}{4}kn}, \quad 0 \leq n < 4 \tag{3.115}$$

we obtain

$$x[n] = \frac{1}{4} \left(\underbrace{X[0]}_{3} e^{j0} + \underbrace{X[1]}_{2+j} e^{j\frac{2\pi}{4}1n} + \underbrace{X[2]}_{-3} e^{j\frac{2\pi}{4}2n} + \underbrace{X[3]}_{2-j} e^{j\frac{2\pi}{4}3n} \right). \tag{3.115}$$

When (3.115) is simplified, we get

$$x[n] = \frac{1}{4}\left(3 + (2+j)e^{j\frac{\pi}{2}n} - 3e^{j\pi n} + (2-j)e^{j\frac{3\pi}{2}n}\right) \qquad (3.116)$$

Evaluating (3.116), i.e., $x[n]$, for $n = 0, 1, 2, 3$, we obtain,

$$x[0] = 1 \quad x[1] = 1 \quad x[2] = -1 \quad x[3] = 2$$

which can be written in short as

$$x[n] = \begin{bmatrix} 1 & 1 & -1 & 2 \end{bmatrix}.$$

Note: Remember that $e^{j\theta} = \cos(\theta) + j\sin(\theta)$.

Example 3.20 Find the discrete Fourier transform of the signal shown in Fig. 3.20.

Solution 3.20 Using the DFT formula

$$X[k] = \sum_{n,N} x[n]e^{-j\frac{2\pi}{N}kn} \quad k, N$$

for $N = 3$, we obtain

$$X[k] = \sum_{n=-1}^{1} x[n]e^{-j\frac{2\pi}{N}kn} \quad -1 \le k \le 1. \qquad (3.117)$$

When (3.117) is expanded, we get

$$X[k] = \underbrace{x[-1]}_{1/2} e^{-j\frac{2\pi}{3}k(-1)} + \underbrace{x[1]}_{1/2} e^{-j\frac{2\pi}{3}k1}, \quad -1 \le k \le 1 \qquad (3.118)$$

which is simplified as

$$X[k] = \cos\left(\frac{2\pi}{3}k\right), \quad -1 \le k \le 1. \qquad (3.119)$$

Fig. 3.20 Aperiodic signal for Example 3.20

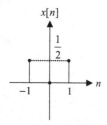

From (3.115) DFT coefficients can be calculated as

$$k = -1 \rightarrow X[-1] = -\frac{1}{2}$$
$$k = 0 \rightarrow X[0] = 1$$
$$k = 1 \rightarrow X[1] = -\frac{1}{2}$$

That is,

$$X[k] = [\; -\tfrac{1}{2} \quad \underbrace{1}_{k=0} \quad -\tfrac{1}{2}\;].$$

Note: For the previous example, discrete Fourier transform is calculated for $N = 3$ which is equal to the length of the aperiodic sequence $x[n]$. Hence, if it is not clearly mentioned, the default length of the DFT computation is the same as the length of the aperiodic sequence $x[n]$.

Example 3.21 DFT coefficients of an aperiodic signal are given as

$$X[k] = [\; -\tfrac{1}{2} \quad \underbrace{1}_{k=0} \quad -\tfrac{1}{2}\;]. \tag{3.120}$$

Find $x[n]$ whose DFT coefficients are $X[k]$.

Solution 3.21 If we use inverse DFT formula

$$x[n] = \frac{1}{N} \sum_{k,N} X[k] e^{j\frac{2\pi}{N}kn}, \quad n, N$$

for the given signal, we get

$$x[n] = \frac{1}{3} \sum_{k=-1}^{1} X[k] e^{j\frac{2\pi}{3}kn}, \quad -1 \leq n \leq 1. \tag{3.121}$$

When the summation term in (3.121) is expanded, we obtain

$$x[n] = \frac{1}{3} \left(\underbrace{X[-1]}_{-1/2} e^{-j\frac{2\pi}{3}n} + \underbrace{X[0]}_{1} e^0 + \underbrace{X[1]}_{-1/2} e^{j\frac{2\pi}{3}n} \right)$$

which is simplified as

$$x[n] = \frac{1}{3}\left(-\frac{1}{2}e^{-j\frac{2\pi}{3}n} + 1 - \frac{1}{2}e^{j\frac{2\pi}{3}n}\right). \tag{3.122}$$

Let's evaluate (3.122), i.e., $x[n]$, for $n = -1, 0, 1$. We first calculate for $n = -1$ as

$$x[-1] = \frac{1}{3}\left(-\frac{1}{2}e^{-j\frac{2\pi}{3}(-1)} + 1 - \frac{1}{2}e^{j\frac{2\pi}{3}(-1)}\right)$$

which is simplified as

$$x[-1] = \frac{1}{3}\left(-\cos(\frac{2\pi}{3}) + 1\right) \rightarrow x[-1] = \frac{1}{3}\left(\frac{1}{2} + 1\right) \rightarrow x[-1] = \frac{1}{2}$$

and for $n = 0$, we have

$$x[n] = \frac{1}{3}\left(-\frac{1}{2}e^0 + 1 - \frac{1}{2}e^0\right) \rightarrow x[n] = 0$$

and finally for $n = 1$, we get

$$x[1] = \frac{1}{3}\left(-\frac{1}{2}e^{-j\frac{2\pi}{3}(1)} + 1 - \frac{1}{2}e^{j\frac{2\pi}{3}(1)}\right)$$

which is simplified as

$$x[1] = \frac{1}{3}\left(-\cos(\frac{2\pi}{3}) + 1\right) \rightarrow x[1] = \frac{1}{3}\left(\frac{1}{2} + 1\right) \rightarrow x[1] = \frac{1}{2}.$$

Thus the signal $x[n]$ has the values

$$x[-1] = \tfrac{1}{2} \quad x[0] = 0 \quad x[1] = \tfrac{1}{2}$$

which is written in more compact form as

$$x[n] = [\tfrac{1}{2} \underbrace{0}_{n=0} \tfrac{1}{2}]. \tag{3.123}$$

Question: For the previous example if we evaluate

$$x[n] = \frac{1}{3}\left(-\frac{1}{2}e^{-j\frac{2\pi}{3}n} + 1 - \frac{1}{2}e^{j\frac{2\pi}{3}n}\right)$$

Fig. 3.21 Aperiodic signal
for Example 3.22

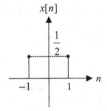

for $n = 0, 1$, and 2, we obtain

$$x[n] = [\underbrace{0}_{n=0} \ \tfrac{1}{2} \ \tfrac{1}{2}]. \tag{3.124}$$

When (3.123) and (3.124) are compared to each other, we see that (3.124) can be obtained from (3.123) by rotate left or rotate right operations.

Example 3.22 Find the 8-point discrete Fourier transform of the signal in Fig. 3.21.

Solution 3.22 Although the length of the aperiodic signal equals to 2, the DFT will be calculated for 8-points. For this reason, we first pad the signal by zeros so that its length equals to 8. So the finite length signal becomes as

$$x[n] = [-1 \ \underbrace{0}_{n=0} \ 1 \ 0 \ 0 \ 0 \ 0 \ 0].$$

And the 8-point DFT is computed using

$$X[k] = \sum_{n=-1}^{6} x[n] e^{-j\frac{2\pi}{8}kn}, \quad -1 \le k \le 6. \tag{3.125}$$

When the summation in (3.125) is expanded, we get

$$X[k] = \left(\frac{1}{2}\right) \times e^{-j\frac{2\pi}{8}k(-1)} + 0 \times e^{-j\frac{2\pi}{8}k0} + \left(\frac{1}{2}\right) \times e^{-j\frac{2\pi}{8}k1} + 0 \times e^{-j\frac{2\pi}{8}k2}$$
$$+ 0 \times e^{-j\frac{2\pi}{8}k3} + 0 \times e^{-j\frac{2\pi}{8}k4} + 0 \times e^{-j\frac{2\pi}{8}k5} + 0 \times e^{-j\frac{2\pi}{8}k6}$$

which is simplified as

$$X[k] = \frac{1}{2}\left(e^{j\frac{2\pi}{8}k} + e^{-j\frac{2\pi}{8}kn}\right). \tag{3.126}$$

Equation (3.126) can be written in terms of $\cos(\cdot)$ function as

$$X[k] = \cos\left(\frac{2\pi}{8}k\right), \quad -1 \le k \le 6. \tag{3.127}$$

And when the Fourier series coefficients in (3.127) are explicitly calculated, we obtain

$$X[k] = \left[\cos\left(-\frac{2\pi}{8}\right) \quad \cos(0) \quad \cos\left(\frac{2\pi}{8}\right) \quad \cos\left(\frac{4\pi}{8}\right) \quad \cos\left(\frac{6\pi}{8}\right) \quad \cos\left(\frac{8\pi}{8}\right) \quad \cos\left(\frac{10\pi}{8}\right) \quad \cos\left(\frac{12\pi}{8}\right)\right]$$

which is simplified as

$$X[k] = [0.7071 \quad \underbrace{1}_{k=0} \quad 0.7071 \quad 0 \quad -0.7071 \quad -1 \quad -0.7071 \quad 0].$$

Example 3.23 DFT coefficients are complex numbers. And those complex coefficients have magnitude and phase values. For the DFT coefficients

$$X[k] = [3 \quad 2+j \quad -3+j \quad 2-j]$$

find $|X[k]|$, i.e., magnitudes of the DFT coefficients, and $\angle X[k]$, i.e., phase information of DFT coefficients.

Solution 3.23 For the complex number $x = a + bj$ the magnitude and phase information is calculated as

$$|x| = \sqrt{(a^2 + b^2)}, \quad \angle \tan^{-1}\left(\frac{b}{a}\right). \tag{3.128}$$

Using (3.128) the magnitude and phase of each DFT coefficient is calculated as

$$|X[0]| = \sqrt{3^2 + 0^2} \to 3 \qquad \angle X[0] = \tan^{-1}\frac{0}{3} \to 0$$
$$|X[1]| = \sqrt{2^2 + 1^2} \to \sqrt{5} \qquad \angle X[1] = \tan^{-1}\frac{1}{2} \to 0.15\pi$$
$$|X[2]| = \sqrt{(-3)^2 + 1^2} \to \sqrt{10} \quad \angle X[2] = \tan^{-1}-\frac{1}{3} \to -0.1\pi$$
$$|X[3]| = \sqrt{2^2 + (-1)^2} \to \sqrt{5} \quad \angle X[3] = \tan^{-1}-\frac{1}{2} \to -0.15\pi$$

Magnitude and phase values are plotted in Fig. 3.22.

Example 3.24 One period of the discrete time Fourier transform of the non-periodic signal $x[n]$ is given in Fig. 3.23. Using the given Fourier transform graph:

(a) Find the 4-point DFT of $x[n]$. (b) Find the 8-point DFT of $x[n]$. (c) Find the 16-point DFT of $x[n]$.

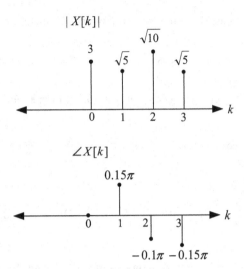

Fig. 3.22 Magnitude and phase plot of DFT coefficients in Example 3.23

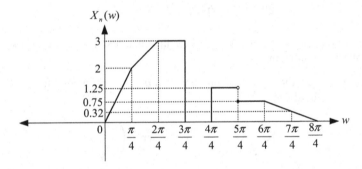

Fig. 3.23 One period of the discrete time Fourier transform of a non-periodic signal

Solution 3.24

(a) DFT coefficients are obtained by sampling of $X_n(w)$ in frequency domain. That
   is,

$$X[k] = X_n(w)|_{w=kw_s} \quad w_s = \frac{2\pi}{N}. \tag{3.129}$$

Since $N = 4$, we take 4 samples from one period of $X_n(w)$. The sampling period
is

$$w_s = \frac{2\pi}{8} \rightarrow w_s = \frac{2\pi}{4}.$$

The sampling operation is illustrated in Fig. 3.24.

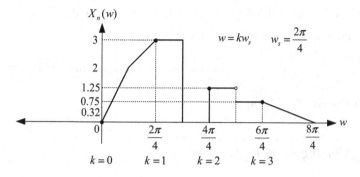

Fig. 3.24 Sampling of the Fourier transform for $N = 4$

Considering Fig. 3.24, the DFT coefficients can be written as

$$X[k] = [0 \quad 3 \quad 1.25 \quad 0.75].$$

(b) For $N = 8$, we take 8 samples from one period of $X_n(w)$. The sampling period is

$$w_s = \frac{2\pi}{8} \rightarrow w_s = \frac{\pi}{4}.$$

The sampling operation for $N = 8$ is illustrated in Fig. 3.25.

Thus the DFT coefficients obtained in Fig. 3.25 can be written as a mathematical sequence as

$$X[k] = [0 \quad 2 \quad 3 \quad 3 \quad 1.25 \quad 0.75 \quad 0.75 \quad 0.32].$$

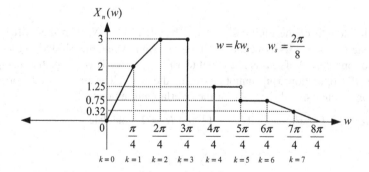

Fig. 3.25 Sampling of the Fourier transform for $N = 8$

Exercise: The aperiodic signal is given as $x[n] = \delta[n] + \delta[n-1]$.

(a) Find the Fourier transform of $x[n]$, i.e., $X_n(w) = ?$
(b) Find $|X_n(w)|$ and $\angle X_n(w)$.
(c) If $\tilde{x}[n] = \sum_{l=-\infty}^{\infty} x[n-4l]$, draw $\tilde{x}[n]$ and using $X_n(w)$, find the Fourier series coefficients of $\tilde{x}[n]$, i.e., $\tilde{X}[k] = ?$
(d) Find 4-point DFT of $x[n]$

3.5.1 Aliasing in Time Domain

When we study sampling theorem, we have seen that during sampling operation if we do not take sufficient number of samples from analog signal, we cannot perfectly reconstruct analog signal at the receiver side from its digital samples. And the effect of this situation is seen as aliasing or overlapping in frequency domain.

We have seen that DFT coefficients of a non-periodic digital signal $x[n]$ are nothing but the samples taken from one period of its Fourier transform, for instance, samples taken for $0 \leq w < 2\pi$. We can reconstruct the digital signal $x[n]$ from its DFT coefficients using

$$x_r[n] = \frac{1}{N} \sum_{k=0}^{N-1} X[k] e^{j\frac{2\pi}{N}kn}, \quad 0 \leq n < N. \tag{3.130}$$

Now we ask the question: Is $x_r[n]$ always equal to $x[n]$? If not always, then what is the criteria for $x_r[n]$ to be equal to $x[n]$?

We know that N-point DFT coeffcients of $x[n]$ equals to the one period of the DFS coefficients of the periodic signal $\tilde{x}[n]$, and the relation between $x[n]$ and $\tilde{x}[n]$ can be stated as

$$\tilde{x}[n] = \sum_{k=-\infty}^{\infty} x[n-kN]. \tag{3.131}$$

Let the length of the digital signal $x[n]$ be M. If $M > N$, then the shifted successor signals $x[n-kN]$ overlap each other. And when the shifted signals are summed, one period of $\tilde{x}[n]$ is not equal to $x[n]$ anymore. This means that using the inverse DFT operation, $x[n]$ cannot be obtained exactly. The amount of distortion in the reconstructed signal depends on the overlapping amount.

Example 3.25 For $x[n] = \begin{bmatrix} -1 & 1 & 1 \end{bmatrix}$ and $N = 2$, calculate

$$\tilde{x}[n] = \sum_{k=-\infty}^{\infty} x[n-kN].$$

Find one period of $\tilde{x}[n]$ and compare it to $x[n]$.

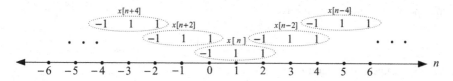

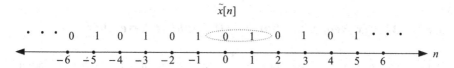

Fig. 3.26 Shifted signals

$$\tilde{x}[n]$$

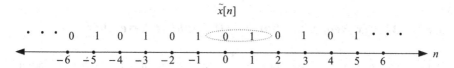

Fig. 3.27 Sum of the shifted signals in Fig. 3.26

Solution 3.25 The shifted signals are shown in Fig. 3.26.

The sum of the shifted signals in Fig. 3.26 yields the signal in Fig. 3.27.

As it is seen from Fig. 3.27, one period of $\tilde{x}[n]$ is $[0\ 1]$ which is totally different than $x[n] = [-1\ \ 1\ \ 1]$.

Example 3.26 $x[n] = [-1\ \ 1\ \ 1]$, calculate 2-point DFT of $x[n]$ and using 2-point DFT coefficients, calculate $x[n]$ using the inverse DFT formula and comment on the results.

Solution 3.26 2-point DFT coefficients of $x[n] = [-1\ \ 1\ \ 1]$ can be calculated using

$$X[k] = \sum_{n=0}^{1} x[n] e^{-j\frac{2\pi}{2}kn}, \quad 0 \le k \le 1$$

yielding

$$X_2[k] = [0\ \ -2].$$

and proceeding in a similar manner 3-point DFT coefficients can be found as

$$X_3[k] = [1\ \ -2\ \ 2].$$

If we use the 2-point inverse DFT formula for $X_2[k]$

$$x[n] = \frac{1}{2} \sum_{k=0}^{1} X_2[k] e^{j\frac{2\pi}{2}kn}, \quad 0 \le n \le 1$$

the aperiodic signal is found as

$$x[n] = \begin{bmatrix} -1 & 1 \end{bmatrix}.$$

which is truncated version of

$$x[n] = \begin{bmatrix} -1 & 1 & 1 \end{bmatrix}.$$

3.5.2   Matrix Representation of DFT and Inverse DFT

Before generalizing the concept, let's consider 3-point DFT of an aperiodic sequence

$$X[k] = \sum_{n=0}^{2} x[n]e^{-j\frac{2\pi}{3}kn}, \quad 0 \le k \le 2 \tag{3.132}$$

When the summation in (3.132) is expanded for each k value, we obtain the following equations

$$\begin{aligned}
X[0] &= x[0]e^0 + x[1]e^0 + x[2]e^0 \\
X[1] &= x[0]e^0 + x[1]e^{-j\frac{2\pi}{3}} + x[2]e^{-j\frac{4\pi}{3}} \\
X[2] &= x[0]e^0 + x[1]e^{-j\frac{4\pi}{3}} + x[2]e^{-j\frac{8\pi}{3}}.
\end{aligned} \tag{3.133}$$

The equation set in (3.133) can be written as

$$\begin{bmatrix} X[0] \\ X[1] \\ X[2] \end{bmatrix} = \begin{bmatrix} x[0] & x[1] & x[2] \end{bmatrix} \times \begin{bmatrix} e^0 & e^0 & e^0 \\ e^0 & e^{-j\frac{2\pi}{3}} & e^{-j\frac{4\pi}{3}} \\ e^0 & e^{-j\frac{4\pi}{3}} & e^{-j\frac{8\pi}{3}} \end{bmatrix}$$

which can be expressed in short as

$$\bar{X}[k] = \bar{x}[n] \times E_N, \quad N = 3. \tag{3.134}$$

From (3.134) $\bar{x}[n]$ can be written as

$$\bar{x}[n] = \bar{X}[k] \times E_N^{-1}. \tag{3.135}$$

In a similar manner, the inverse 3-point DFT formula can be written in matrix form. Expanding

$$x[n] = \frac{1}{3}\sum_{k=0}^{2} X[k]e^{j\frac{2\pi}{3}kn}, \quad 0 \le n \le 2 \tag{3.136}$$

we get

$$
\begin{aligned}
x[0] &= \frac{1}{3}\left(X[0]e^0 + X[1]e^0 + X[2]e^0\right)\\
x[1] &= \frac{1}{3}\left(X[0]e^0 + X[1]e^{j\frac{2\pi}{3}} + X[2]e^{j\frac{4\pi}{3}}\right)\\
x[2] &= \frac{1}{3}\left(X[0]e^0 + X[1]e^{j\frac{4\pi}{3}} + X[2]e^{j\frac{8\pi}{3}}\right).
\end{aligned}
\tag{3.137}
$$

The equation set in (3.137) can be written in matrix form as

$$
\begin{bmatrix} x[0] \\ x[1] \\ x[2] \end{bmatrix} = \frac{1}{3} \times [X[0] \quad X[1] \quad X[2]] \times \begin{bmatrix} e^0 & e^0 & e^0 \\ e^0 & e^{j\frac{2\pi}{3}} & e^{j\frac{4\pi}{3}} \\ e^0 & e^{j\frac{4\pi}{3}} & e^{j\frac{8\pi}{3}} \end{bmatrix}. \tag{3.138}
$$

When (3.138) is compared to (3.139)

$$\bar{x}[n] = \bar{X}[k] \times E_N^{-1} \tag{3.139}$$

we obtain

$$E_N^{-1} = \frac{1}{N}E_N^*. \tag{3.140}$$

Note: E_N^* is the conjugate of E_N. If $e = a + jb$ then conjugate of e is $e^* = a - jb$ and if $e = e^{j\theta}$ then $e^* = e^{-j\theta}$.

3.5.3 Properties of the Discrete Fourier Transform

Since there is a close relationship between discrete Fourier series coefficients of a periodic signal and the discrete Fourier transform of its one period, it is logical to review the properties of the discrete Fourier series coefficients of a periodic signal.

For the three periodic signals

$$
\begin{aligned}
\tilde{x}[n] &\rightarrow Periodic\,with\,period\,N\\
\tilde{x}_1[n] &\rightarrow Periodic\,with\,period\,N\\
\tilde{x}_2[n] &\rightarrow Periodic\,with\,period\,N
\end{aligned}
$$

let's denote the Fourier series coefficients by

$$\tilde{X}[k] \rightarrow Periodic\,with\,period\,N$$
$$\tilde{X}_1[k] \rightarrow Periodic\,with\,period\,N$$
$$\tilde{X}_2[k] \rightarrow Periodic\,with\,period\,N.$$

And the correspondence between signals and their DFS coefficients are shown as

$$\tilde{x}[n] \overset{\text{DFS}}{\leftrightarrow} \tilde{X}[k]$$
$$\tilde{x}_1[n] \overset{\text{DFS}}{\leftrightarrow} \tilde{X}_1[k]$$
$$\tilde{x}_2[n] \overset{\text{DFS}}{\leftrightarrow} \tilde{X}_2[k].$$

Properties
Linearity:

$$a\tilde{x}_1[n] + b\tilde{x}_2[n] \overset{\text{DFS}}{\leftrightarrow} a\tilde{X}_1[k] + b\tilde{X}_2[k]$$

Duality:

$$\tilde{X}[n] \overset{\text{DFS}}{\leftrightarrow} N\tilde{x}[-k]$$

Shifting in time:

$$\tilde{x}[n - m] \overset{\text{DFS}}{\leftrightarrow} e^{-j\frac{2\pi}{N}km}\tilde{X}[k]$$

Shifting in frequency:

$$e^{j\frac{2\pi}{N}ln}\tilde{x}[n] \overset{\text{DFS}}{\leftrightarrow} \tilde{X}[k - l]$$

Convolution in time domain:

$$\sum_{m=0}^{N-1} \tilde{x}_1[m]\tilde{x}_2[n - m] \overset{\text{DFS}}{\leftrightarrow} \tilde{X}_1[k]\tilde{X}_2[k]$$

Convolution in frequency domain:

$$\tilde{x}_1[n]\tilde{x}_2[n] \overset{\text{DFS}}{\leftrightarrow} \frac{1}{N}\sum_{k=0}^{N-1} \tilde{X}_1[m]\tilde{X}_2[k - m]$$

Conjugate:

$$\tilde{x}^*[-n] \overset{\text{DFS}}{\leftrightarrow} \tilde{X}^*[k]$$

Real part DFS:

$$Re\{\tilde{x}[n]\} \overset{\text{DFS}}{\leftrightarrow} \frac{1}{2}\left(\tilde{X}[k] + \tilde{X}^*[-k]\right)$$

Imaginary part DFS:

$$jIm\{\tilde{x}[n]\} \overset{\text{DFS}}{\leftrightarrow} \frac{1}{2}\left(\tilde{X}[k] - \tilde{X}^*[-k]\right)$$

Real part:

$$\frac{1}{2}\left(\tilde{x}[n] + \tilde{x}^*[-n]\right) \overset{\text{DFS}}{\leftrightarrow} Re\{\tilde{X}[k]\}$$

Imaginary part:

$$\frac{1}{2}\left(\tilde{x}[n] - \tilde{x}^*[-n]\right) \overset{\text{DFS}}{\leftrightarrow} jIm\{\tilde{X}[k]\}$$

For real $\tilde{x}[n]$, we have the following properties
Conjugate:

$$\tilde{X}[k] = \tilde{X}^*[-k]$$

Real DFT coefficients:

$$Re\{\tilde{X}[k]\} = Re\{\tilde{X}^*[-k]\}$$

Imaginary DFT coefficients:

$$Im\{\tilde{X}[k]\} = -Im\{\tilde{X}^*[-k]\}$$

Absolute value:

$$\left|\tilde{X}[k]\right| = \left|\tilde{X}[-k]\right|$$

Phase value:

$$\angle\tilde{X}[k] = -\angle\tilde{X}[-k]$$

Real part:

$$\frac{1}{2}\left(\tilde{x}[n] + \tilde{x}[-n]\right) \overset{DFS}{\longleftrightarrow} Re\{\tilde{X}[k]\}$$

Imaginary part:

$$\frac{1}{2}\left(\tilde{x}[n] - \tilde{x}[-n]\right) \overset{DFS}{\longleftrightarrow} jIm\{\tilde{X}[k]\}$$

Note: If $x[n] = a[n] + jb[n]$, then $x^*[n] = a[n] - jb[n]$

3.5.4   Circular Convolution

The discrete Fourier transform of an aperiodic sequence $x[n]$ with length N equals to the one period of the Fourier series coefficients of the periodic signal $\tilde{x}[n]$ obtained from $x[n]$ as

$$\tilde{x}[n] = \sum_{k=-\infty}^{\infty} x[n - kN]$$

and the relation between DFT coefficients of $x[n]$ and one period of Fourier series coefficients of the periodic signal $\tilde{x}[n]$ is given as

$$X[k] = \begin{cases} \tilde{X}[k] & \text{if } 0 \leq k \leq N-1 \\ 0 & \text{otherwise.} \end{cases}$$

Let's denote one period of $\tilde{x}[n]$ for $0 \leq n \leq N - 1$ by $x[(n)_N]$. It is clear that if the length of $x[n]$ is N then $x[(n)_N] = x[n]$. However, if the length of $x[n]$ is a number other than N then

$$x[(n)_N] \neq x[n].$$

If not indicated otherwise, we will assume that the length of $x[n]$ and period of $\tilde{x}[n]$ are equal to each other.

Properties

$$x_1[n] \rightarrow Aperiodic\ signal\ with\ length\ N_1$$
$$x_2[n] \rightarrow Aperiodic\ signal\ with\ length\ N_2$$

$$N = \max\{N_1, N_2\}$$

$$x_1[n] \overset{\text{N-pointDFT}}{\longleftrightarrow} X_1[k]$$

$$x_2[n] \overset{\text{N-pointDFT}}{\longleftrightarrow} X_2[k]$$

Linearity:

$$ax_1[n] + bx_2[n] \overset{\text{DFT}}{\longleftrightarrow} aX_1[k] + aX_2[k]$$

Circular Shifting:

$$x\left[(n-m)_N\right] \overset{\text{DFT}}{\longleftrightarrow} e^{-j\frac{2\pi}{N}km}X[k]$$

Duality:

$$x[n] \overset{\text{DFT}}{\longleftrightarrow} X[k]$$

$$X[n] \overset{\text{DFT}}{\longleftrightarrow} Nx\left[(-k)_N\right]$$

Symmetry:

$$x^*[n] \overset{\text{DFT}}{\longleftrightarrow} X^*\left[(-k)_N\right]$$

$$X^*\left[(-n)_N\right] \overset{\text{DFT}}{\longleftrightarrow} X^*[k]$$

Symmetry property leads to the following properties

$$Re\{x[n]\} \overset{\text{DFT}}{\longleftrightarrow} X_{ep}[k], \quad ep:even\,part$$

$$jIm\{x[n]\} \overset{\text{DFT}}{\longleftrightarrow} X_{op}[k], \quad op:odd\,part$$

$$x_{ep}[n] \overset{\text{DFT}}{\longleftrightarrow} Re\{X[k]\}$$

$$x_{op}[n] \overset{\text{DFT}}{\longleftrightarrow} jIm\{X[k]\}$$

Circular Convolution:

$$x_1[n] \overset{\text{DFT}}{\longleftrightarrow} X_1[k]$$
$$x_2[n] \overset{\text{DFT}}{\longleftrightarrow} X_2[k]$$

If

$$Y[k] = X_1[k]X_2[k]$$

then

$$y[n] = \sum_{m=0}^{N-1} x_1[m]x_2[(n-m)_N]$$

or

$$y[n] = \sum_{m=0}^{N-1} x_2[m]x_1[(n-m)_N].$$

And the expression

$$\sum_{m=0}^{N-1} x_1[m]x_2[(n-m)_N]$$

is called the circular convolution of $x_1[n]$ and $x_2[n]$ and denoted by

$$x_1[n] \; Ⓝ \; x_2[n].$$

Example 3.27 What does $x[(-n)_5] \quad 0 \le n \le 4$ mean?

Solution 3.27 $x[(-n)_5]$ equals to one period of $\tilde{x}[-n]$ in the interval $0 \le n \le 4$, i.e.,

$$x[(-n)_5] = \tilde{x}[-n] \quad 0 \le n \le 4$$

and

$$\tilde{x}[n] = \sum_{l=-\infty}^{\infty} x[n-5l].$$

Note: We assumed that the length of $x[n]$ and period of $\tilde{x}[n]$ are equal to each other.

Example 3.28 If $x[n] = [-1 \quad 1 \quad -1 \quad 0.5 \quad -1]$, find $x[(-n)_5] \; 0 \le n \le 4$.

Solution 3.28 $x[(-n)_5]$ equals to $\tilde{y}[n] = \tilde{x}[-n]$ for $0 \le n \le 4$ and $\tilde{x}[n]$ is given as

$$\tilde{x}[n] = \sum_{l=-\infty}^{\infty} x[n - 5l].$$

One period of $\tilde{x}[-n]$ in the interval $0 \leq n \leq 4$ is found by employing rotate inside operation on one period of $\tilde{x}[n]$, i.e., on $x[n]$. That is

$$x\big[(-n)_5\big] = RI(x[n])$$

which can be calculated as

$$x\big[(-n)_5\big] = [-1 \quad 1 \quad 0.5 \quad -1 \quad 1].$$

Example 3.29 If $x[n] = [-1 \quad 1 \quad -1 \quad 0.5 \quad -1]$, find $x\big[(1 - n)_5\big]$.

Solution 3.29 $x\big[(1 - n)_5\big]$ equals to $\tilde{x}[1 - n]$ for $0 \leq n \leq 4$ and $\tilde{x}[n]$ is calculated as

$$\tilde{x}[n] = \sum_{l=-\infty}^{\infty} x[n - 5l].$$

One period of $\tilde{x}[1 - n]$ is obtained by rotating one period of $\tilde{x}[-n]$ to the right by '1' unit. That is

$$x\big[(1 - n)_5\big] = RR\big(x\big[(-n)_5\big]\big).$$

Using the result of the previous example, i.e.,

$$x\big[(-n)_5\big] = [-1 \quad -1 \quad 0.5 \quad -1 \quad 1]$$

we can calculate $x\big[(1 - n)_5\big]$ via

$$x\big[(1 - n)_5\big] = RR\big(x\big[(-n)_5\big]\big)$$

which yields

$$x\big[(1 - n)_5\big] = [1 \quad -1 \quad -1 \quad 0.5 \quad -1].$$

Note: $x\big[(2 - n)_5\big]$ is obtained by rotating $x\big[(1 - n)_5\big]$ to the right by '1' unit. And $x\big[(-1 - n)_5\big]$ is obtained by rotating $x\big[(-n)_5\big]$ to the left by '1' unit.

Example 3.30 If $x[n] = [-1 \quad -1 \quad 1 \quad 1]$, find $x\big[(-n)_3\big]$.

Solution 3.30 $x\big[(-n)_3\big] = \tilde{x}[-n]$ for $0 \leq n \leq 3$ and $\tilde{x}[n]$ is obtained as

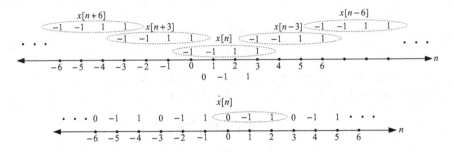

Fig. 3.28 Shifted replicas of $x[n]$ and calculation of $\tilde{x}[n]$

$$\tilde{x}[n] = \sum_{l=-\infty}^{\infty} x[n - 3l].$$

Since the length of $x[n]$ is 4, the shifted successor copies in $\sum_{l=-\infty}^{\infty} x[n - 3l]$ overlap with each other. For this reason, one period of $\tilde{x}[n]$ is not equal to $x[n]$ anymore. It should be calculated explicitly. This calculation is explained in Fig. 3.28.

$x\big[(n)_3\big]$ for $0 \leq n \leq 3$ equals to one period of $\tilde{x}[n]$ and from Fig. 3.28, it is found as

$$\tilde{x}_{op}[n] = \begin{bmatrix} 0 & -1 & 1 \end{bmatrix}$$

which is denoted by $x\big[(n)_3\big]$, that is,

$$x\big[(n)_3\big] = \begin{bmatrix} 0 & -1 & 1 \end{bmatrix}.$$

And $x\big[(-n)_3\big]$ which is equal to one period of $\tilde{x}[-n]$ can be found using the rotate inside operation as

$$x\big[(-n)_3\big] = RI\big(\tilde{x}_{op}[n]\big)$$

yielding

$$x\big[(-n)_3\big] = \begin{bmatrix} 0 & 1 & -1 \end{bmatrix}.$$

Exercise: For the previous example find $x\big[(2 - n)_3\big]$.

Example 3.31 If $x[n] = \begin{bmatrix} 0.5 & 0.5 & -0.5 & 1 & -1 \end{bmatrix}$, find $x[(n - 2)_5]$.

Solution 3.31 $x[(n - 2)_5]$ equals to $\tilde{x}[n - 2]$ for $0 \leq n \leq 4$ and $\tilde{x}[n]$ is obtained as

$$\tilde{x}[n] = \sum_{l=-\infty}^{\infty} x[n - lN]$$

where $N = 5$. Since the length of $x[n]$ equals to the period value of the $\tilde{x}[n]$, then $\tilde{x}[n]$ in one period interval $0 \leq n \leq 4$ equals to $x[n]$. And one period of the shifted periodic signal for $0 \leq n \leq 4$ can be obtained by rotate right operation as

$$\tilde{x}_{op}[n - 2] = RR(x[n], 2)$$

which can be calculated in two steps as follows

$$\tilde{x}_{op}[n - 1] = RR(x[n], 1)$$
$$= \begin{bmatrix} -1 & 0.5 & 0.5 & -0.5 & 1 \end{bmatrix}$$

$$\tilde{x}_{op}[n - 2] = RR(\tilde{x}_{op}[n - 1], 1)$$
$$= \begin{bmatrix} 1 & -1 & 0.5 & 0.5 & -0.5 \end{bmatrix}.$$

As a result $x[(n - 2)_5]$ is found as

$$x[(n - 2)_5] = \begin{bmatrix} 1 & -1 & 0.5 & 0.5 & -0.5 \end{bmatrix}.$$

Example 3.32 If $x_1[n] = \begin{bmatrix} -1 & -1 & 1 & 0.5 \end{bmatrix}$ and $x_2[n] = \begin{bmatrix} -1 & -1 & -1 & 1 \end{bmatrix}$, find 4-point circular convolution of $x_1[n]$ and $x_2[n]$. That is,

$$x_1[n] \ \textcircled{4} \ x_2[n] = ?$$

Solution 3.32 Method 1: N-point circular convolution of $x_1[n]$ and $x_2[n]$ can be calculated using

$$x_1[n] \ \textcircled{N} \ x_2[n] = \sum_{m=0}^{N-1} x_1[m]x_2[(n - m)_N]. \qquad (3.141)$$

Let $y[n] = x_1[n] \ \textcircled{N} \ x_2[n]$, expanding the right hand side of (3.141) for $N = 4$ we get

$$y[n] = x_1[0]x_2[(n)_4] + x_1[1]x_2[(n - 1)_4] + x_1[2]x_2[(n - 2)_4]$$
$$+ x_1[3]x_2[(n - 3)_4] \qquad (3.142)$$

where the signals $x_2[(n)_4]$, $x_2[(n - 1)_4]$, $x_2[(n - 2)_4]$, and $x_2[(n - 3)_4]$ can be calculated as

$$x_2\big[(n)_4\big] = x_2[n] \rightarrow x_2\big[(n)_4\big] = \begin{bmatrix} 1 & -1 & -1 & 1 \end{bmatrix}$$
$$x_2\big[(n-1)_4\big] = RR(x_2[n], 1) \rightarrow x_2\big[(n-1)_4\big] = \begin{bmatrix} 1 & 1 & -1 & -1 \end{bmatrix}$$
$$x_2\big[(n-2)_4\big] = RR(x_2[n], 2) \rightarrow x_2\big[(n-2)_4\big] = \begin{bmatrix} -1 & 1 & 1 & -1 \end{bmatrix} \tag{3.143}$$
$$x_2\big[(n-3)_4\big] = RR(x_2[n], 3) \rightarrow x_2\big[(n-3)_4\big] = \begin{bmatrix} -1 & -1 & 1 & 1 \end{bmatrix}.$$

Substituting the calculated values in (3.143) into (3.142), we get

$$y[n] = (-1) \times \begin{bmatrix} 1 & -1 & -1 & 1 \end{bmatrix} + (-1) \times \begin{bmatrix} 1 & 1 & -1 & -1 \end{bmatrix}$$
$$+ (1) \times \begin{bmatrix} -1 & 1 & 1 & -1 \end{bmatrix} + (0.5) \times \begin{bmatrix} -1 & -1 & 1 & 1 \end{bmatrix}$$

which is simplified as

$$y[n] = \begin{bmatrix} -3.5 & 0.5 & 3.5 & -0.5 \end{bmatrix}.$$

Method 2: N-point circular convolution of $x_1[n]$ and $x_2[n]$ can be calculated as

$$y[n] = \sum_{m=0}^{N-1} x_1[m] x_2\big[(n-m)_N\big]. \tag{3.144}$$

Evaluating the right hand side of (3.144) for the n values in the range $0 \leq n \leq N-1$, we get the equation set

$$y[0] = \sum_{m=0}^{N-1} x_1[m] x_2\big[(0-m)_N\big]$$
$$y[1] = \sum_{m=0}^{N-1} x_1[m] x_2\big[(1-m)_N\big] \tag{3.145}$$
$$\vdots$$
$$y[N-1] = \sum_{m=0}^{N-1} x_1[m] x_2\big[(N-1-m)_N\big].$$

For $N = 4$ equation set (3.145) becomes as

$$y[0] = \sum_{m=0}^{3} x_1[m]x_2\big[(0-m)_4\big]$$

$$y[1] = \sum_{m=0}^{3} x_1[m]x_2\big[(1-m)_4\big]$$

$$y[2] = \sum_{m=0}^{3} x_1[m]x_2\big[(2-m)_4\big]$$ (3.146)

$$y[3] = \sum_{m=0}^{3} x_1[m]x_2\big[(3-m)_4\big]$$

where the signals $x_2\big[(-m)_4\big]$, $x_2\big[(1-m)_4\big]$, $x_2\big[(2-m)_4\big]$, and $x_2\big[(3-m)_4\big]$ are calculated as

$$x_2\big[(-m)_4\big] = RI(x_2[m]) \rightarrow x_2\big[(-m)_4\big] = \begin{bmatrix} 1 & 1 & -1 & -1 \end{bmatrix}$$
$$x_2\big[(1-m)_4\big] = RR\big(x_2\big[(-m)_4\big], 1\big) \rightarrow x_2\big[(1-m)_4\big] = \begin{bmatrix} -1 & 1 & 1 & -1 \end{bmatrix}$$
$$x_2\big[(2-m)_4\big] = RR\big(x_2\big[(1-m)_4\big], 1\big) \rightarrow x_2\big[(2-m)_4\big] = \begin{bmatrix} -1 & -1 & 1 & 1 \end{bmatrix}$$
$$x_2\big[(3-m)_4\big] = RR\big(x_2\big[(2-m)_4\big], 1\big) \rightarrow x_2\big[(3-m)_4\big] = \begin{bmatrix} 1 & -1 & -1 & 1 \end{bmatrix}.$$

Now consider the summation term

$$y[0] = \sum_{m=0}^{3} x_1[m]x_2\big[(0-m)_4\big].$$ (3.147)

Let $w[m] = x_2\big[(-m)_4\big]$ i.e., $w[m] = \begin{bmatrix} 1 & 1 & -1 & -1 \end{bmatrix}$; then expanding (3.147), we obtain

$$y[0] = x_1[0]w[0] + x_1[1]w[1] + x_1[2]w[2] + x_1[3]w[3]$$

which is nothing but dot product of two vectors $x_1[n]$ and $w[n]$, that is

$$y[0] = \begin{bmatrix} x_1[0] & x_1[1] & x_1[2] & x_1[3] \end{bmatrix} \cdot \begin{bmatrix} w[0] & w[1] & w[2] & w[3] \end{bmatrix}$$

which can also be written as

$$y[0] = x_1[m] \cdot w[m]$$

or

$$y[0] = x_1[m] \cdot x_2\big[(-m)_4\big].$$

Then

$$y[0] = (-1) \times (1) + (-1) \times (1) + (1) \times (-1) + (0.5) \times (-1)$$
$$y[0] = -3.5.$$

In a similar manner,

$$y[1] = x_1[m] \cdot x_2[(1-m)_4]$$
$$y[1] = (-1) \times (-1) + (-1) \times (1) + (1) \times (1) + (0.5) \times (-1)$$
$$y[1] = 0.5$$

$$y[2] = x_1[m] \cdot x_2[(2-m)_4]$$
$$y[2] = (-1) \times (-1) + (-1) \times (-1) + (1) \times (1) + (0.5) \times (1)$$
$$y[2] = 3.5$$

$$y[3] = x_1[m] \cdot x_2[(3-m)_4]$$
$$y[3] = (-1) \times (1) + (-1) \times (-1) + (1) \times (-1) + (0.5) \times (1)$$
$$y[3] = -0.5$$

As a result;

$$y[n] = [-3.5 \quad 0.5 \quad 3.5 \quad -0.5].$$

Note: If $y[n] = x_1[n] \,\circledN\, x_2[n]$ and the length of $x_1[n]$ or $x_2[n]$ is shorter than N then the shorter sequence is padded by zeros so that its length equals to N. If both sequences are shorter than N samples then both sequences are padded by zeros so that their lengths equal to N.

Example 3.33 If $x_1[n] = [-1 \quad -1 \quad 1 \quad 0.5]$ and $x_2[n] = [1 \quad -1 \quad -2]$, find 6-point circular convolution of $x_1[n]$ and $x_2[n]$. That is,

$$x_1[n]\,\circled{6}\,x_2[n] =?$$

Solution 3.33 The lengths of the sequences $x_1[n]$ and $x_2[n]$ are 4 and 3 respectively. Both sequences should be padded by zeros so that their lengths equals to 6. That is,

$$x_1[n] = [-1 \quad -1 \quad 1 \quad 0.5 \quad 0 \quad 0] \quad x_2[n] = [1 \quad -1 \quad -2 \quad 0 \quad 0 \quad 0].$$

Then circular convolution operations can be performed as in Example 3.32.
Matrix Representation of Circular Convolution

Example 3.34 If $x_1[n] = [x_1[0] \quad x_1[1] \quad x_1[2]] \quad x_2[n] = [x_2[0] \quad x_2[1] \quad x_2[2]]$
Express 3-point circular convolution of $x_1[n]$ and $x_2[n]$ as matrix multiplication.

Solution 3.34 Expanding the expression

$$y[n] = \sum_{m=0}^{N-1} x_1[m] x_2 \left[(n - m)_N \right]$$

for $N = 3$, we get

$$y[n] = x_1[0] x_2 \left[(n)_N \right] + x_1[1] x_2 \left[(n - 1)_N \right] + x_1[2] x_2 \left[(n - 2)_N \right]$$

which is calculated as

$$y[n] = x_1[0] \begin{bmatrix} x_2[0] & x_2[1] & x_2[2] \end{bmatrix} + x_1[1] [\, x_2[2] & x_2[0] & x_2[1] \,]$$
$$+ x_1[2] [\, x_2[1] & x_2[2] & x_2[0] \,].$$

The expression in (3.48) can be written using matrix multiplication as

$$\begin{pmatrix} y[0] \\ y[1] \\ y[2] \end{pmatrix} = \begin{pmatrix} x_1[0] & x_1[1] & x_1[2] \end{pmatrix} \times \begin{pmatrix} x_2[0] & x_2[2] & x_2[1] \\ x_2[1] & x_2[0] & x_2[2] \\ x_2[2] & x_2[1] & x_2[0] \end{pmatrix}.$$

Example 3.35 If $x[n] = [-1 \quad 0 \quad \underset{n=0}{\underbrace{1}} \quad -1 \quad 2 \quad 1]$, find

$$x_1[n] \; ⑥ \; x_1[n] = ?$$

Solution 3.35 Since the index $n = 0$ is not at the first element in $x[n]$, it is easier to calculate the circular convolution using the first method we introduced. That is expanding

$$y[n] = \sum_{m=-2}^{3} x_1[m] x_1 \left[(n - m)_N \right]$$

for m values, we obtain

$$y[n] = x_1[-2] x_1 \left[(n + 2)_6 \right] + x_1[-1] x_1 \left[(n + 1)_6 \right] + x_1[0] x_1 \left[(n)_6 \right]$$
$$+ x_1[1] x_1 \left[(n - 1)_6 \right] + x_1[2] x_1 \left[(n - 2)_6 \right] + x_1[3] x_1 \left[(n - 3)_6 \right]$$

and placing the n values in the range $-2 \le n \le 3$ for $y[n]$, we can find the 6-point circular convolution result.

Exercise: Prove the following property

$$x\left[(n-m)_N\right] \overset{\text{DFT}}{\leftrightarrow} e^{-j\frac{2\pi}{N}km}X[k].$$

The Relationship between Circular and Linear Convolution:

$$x_1[n] \rightarrow Aperiodic\ signal\ with\ length\ L$$
$$x_2[n] \rightarrow Aperiodic\ signal\ with\ length\ P$$

Linear convolution of $x_1[n]$ and $x_2[n]$ is calculated using

$$y_{lc}[n] = \sum_{m=-\infty}^{\infty} x_1[m]x_2[n-m].$$

The length of $y_{lc}[n]$ is $L+P-1$. N-point circular convolution of $x_1[n]$ and $x_2[n]$
is

$$y_{cc}[n] = x_1[n] \,Ⓝ\, x_2[n].$$

The relationship between $y_{lc}[n]$ and $y_{cc}[n]$ is given as

$$y_{cc}[n] = \begin{cases} \sum_{r=-\infty}^{\infty} y_{lc}[n-rN] & 0 \le n \le N-1 \\ 0 & otherwise. \end{cases}$$

If $N \ge L+P-1$ then the circular convolution and linear convolution results are
the same, i.e., $y_{lc}[n] = y_{cc}[n]$.

3.6   Practical Calculation of the Linear Convolution

Overlap Add and Overlap Save Methods

For practical communication systems, the input signal may not be of finite duration.
It may be of infinite duration or may be a very long sequence, such as TV signal,
video or speech signal.

The input signal $x[n]$ is usually passed through a filter with an impulse response
$h[n]$. Filtering operation is nothing but the convolution of the input signal with the
impulse response of the filter, the filter output is

$$y[n] = x[n] * h[n].$$

If the input signal is very long, then convolution operation takes too much time,
or sometimes it may not still be possible to evaluate the convolution result.

To overcome this issue, two approaches are followed to evaluate the convolution of a very long input and a short impulse response sequences. These methods are called overlap-add and overlap-save. Let's first explain the overlap-add method.

3.6.1 Evaluation of Convolution Using Overlap-Add Method

Let $x[n]$ be the input signal with length N and $h[n]$ be the filter response with length P such that $N > P$. The overlap-add method to evaluate

$$x[n] * h[n]$$

consists of the following steps:

(1) Divide the input sequence to frames such that each frame has length L.

Let's denote the frames by $x_0[n], x_1[n], x_2[n] \cdots \quad 0 \leq n \leq L-1$

(2) Evaluate the convolution of each frame with $h[n]$, i.e., evaluate

$$y_k[n] = x_k[n] * h[n] \quad k = 0, 1, 2, \ldots$$

(3) Calculate the convolution result as

$$y[n] = \sum_{k=0}^{\infty} y_k[n - Lk].$$

Let's explain overlap-add method with an example.

Example 3.36 If $x[n] = \begin{bmatrix} -1 & 1 & 0 & 1 & -1 & 0 & 1 & 1 & -1 & 1 & 0 & -1 \end{bmatrix}$ and $h[n] = \begin{bmatrix} 1 & -1 \end{bmatrix}$, find $x[n] * h[n]$ using overlap-add method.

Solution 3.36

(1) In step 1, we divide the input sequence into frames of length L. The length of the impulse response $h[n]$ is $P = 2$. The length of the frames depends on our choice. Let's choose the length of the frames as $L = 3$ and divide the sequence $x[n]$ into frames as shown in (3.148)

$$x[n] = [\underbrace{-1 \quad 1 \quad 0}_{x_0[n]} \quad \underbrace{1 \quad -1 \quad 0}_{x_1[n]} \quad \underbrace{1 \quad 1 \quad -1}_{x[n]} \quad \underbrace{1 \quad 0 \quad -1}_{x_3[n]}] \qquad (3.148)$$

If the last frame had a length smaller than 3, then we would pad it by zeros until its length equals to 3. The divided frames are

$$x_0[n] = [-1 \quad 1 \quad 0] \quad x_1[n] = [1 \quad -1 \quad 0]$$
$$x_2[n] = [1 \quad 1 \quad -1] \quad x_3[n] = [1 \quad 0 \quad -1].$$

(3.149)

(2) In step 2, we take the convolution of each frame in (3.149) with impulse response $h[n]$.

Let's first calculate the convolution of $x_0[n]$ and $h[n]$, i.e., calculate $y_0[n] = x_0[n] * h[n]$ which is written as

$$y_0[n] = \sum_{k=-\infty}^{\infty} h[k]x[n-k]$$

(3.150)

When (3.150) is expanded for $n = 0, 1, 2, 3$, we obtain

$$y_0[n] = x_0[n] * h[n] \rightarrow y_0[n] = [-1 \quad 1 \quad 0] * [1 \quad -1] \rightarrow y_0[n] = [-1 \quad -2 \quad -1 \quad 0]$$

$$y_1[n] = x_1[n] * h[n] \rightarrow y_0[n] = [1 \quad -1 \quad 0] * [1 \quad -1] \rightarrow y_1[n] = [-1 \quad -2 \quad 1 \quad 0]$$
$$y_2[n] = x_2[n] * h[n] \rightarrow y_0[n] = [1 \quad 1 \quad -1] * [1 \quad -1] \rightarrow y_2[n] = [1 \quad 2 \quad -1 \quad 0]$$

$$y_3[n] = x_3[n] * h[n] \rightarrow y_0[n] = [1 \quad 0 \quad -1] * [1 \quad -1] \rightarrow y_3[n] = [1 \quad -1 \quad -1 \quad 1].$$

(3.151)

(3) In this step, using the results of (3.151) in

$$y[n] = \sum_{k=0}^{\infty} y_k[n - Lk]$$

Fig. 3.29 Shifting of $y_1[n]$

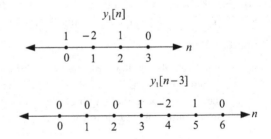

for $L = 3$, we obtain

$$y[n] = \sum_{k=0}^{3} y_k[n - k3]$$

which is expanded as

$$y[n] = y_0[n] + y_1[n - 3] + y_2[n - 6] + y_3[n - 9]. \tag{3.152}$$

The signal $y_1[n - 3]$ in (3.152) is obtained by shifting the amplitudes of $y_1[n]$ to the right by 3 units. When amplitudes are shifted to the right, zero amplitude values are inserted into the old positions.

This means that $y_1[n - 3]$ can be obtained by padding 3 zeros to the beginning of $y_1[n]$. This operation is illustrated in Fig. 3.29.

Thus, the shifted signals together with $y_0[n]$ can be written as

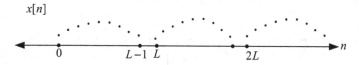

Fig. 3.30 Dividing $x[n]$ into frames

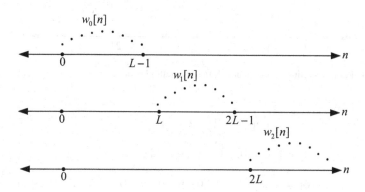

Fig. 3.31 Divided frames of $x[n]$ are shown separately

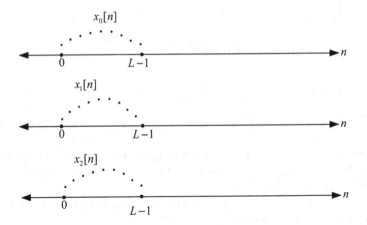

Fig. 3.32 Divided frames of $x[n]$ start at $n = 0$

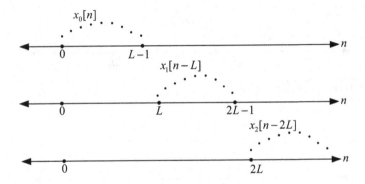

Fig. 3.33 Frames starting at $n = 0$ are shifted by multiples of L

$$y_0[n] = [\underbrace{-1}_{n=0} \quad 2 \quad -1 \quad 0]$$

$$y_1[n-3] = [\underbrace{0}_{n=0} \quad 0 \quad 0 \quad 1 \quad -2 \quad 1 \quad 0]$$

$$y_2[n-6] = [\underbrace{0}_{n=0} \quad 0 \quad 0 \quad 0 \quad 0 \quad 0 \quad 1 \quad 0 \quad -2 \quad 1] \qquad (3.153)$$

$$y_3[n-9] = [\underbrace{0}_{n=0} \quad 0 \quad 0 \quad 0 \quad 0 \quad 0 \quad 0 \quad 0 \quad 1 \quad -1 \quad -1 \quad 1].$$

When the shifted signals in (3.153) are summed, we obtain the convolution result as

$$y[n] = \begin{bmatrix} -1 & 2 & -1 & 1 & -2 & 1 & 1 & 0 & -2 & 2 & -1 & -1 & 1 \end{bmatrix}.$$

Now let's see the mathematical derivation of the overlap-add method.

Assume that the digital sequence $x[n]$ is divided into frames as shown in Fig. 3.30.

And the frames are separately shown in Fig. 3.31.

Let's make the starting index of every frame be equal to $n = 0$. This is shown in Fig. 3.32.

We can obtain the digital signal $x[n]$ by shifting and summing the frames that starts at $n = 0$ as shown in Fig. 3.33.

This operation is mathematically written as

$$x[n] = \sum_{k=0}^{\infty} x_k[n - Lk].$$

Then the convolution of $x[n]$ and $h[n]$ can be written as

$$\begin{aligned} y[n] &= h[n] * x[n] \\ &= h[n] * \sum_{k=0}^{\infty} x_k[n - Lk]. \end{aligned} \tag{3.154}$$

When the summation term in (3.154) is expanded, we get

$$y[n] = h[n] * (x_0[n] + x_1[n - L] + x_2[n - 2L] + \cdots). \tag{3.155}$$

And for linear time invariant systems if

$$y_1[n] = h[n] * x_1[n]$$

then

$$y_1[n - L] = h[n] * x_1[n - L].$$

Using a similar approach for the other convolutional expressions appearing in (3.155), we get

$$y[n] = y_0[n] + y_1[n - L] + y_2[n - 2L] + \cdots$$

where

$$y_0[n] = h[n] * x_0[n], \quad y_1[n] = h[n] * x_1[n], \quad y_2[n] = h[n] * x_2[n].$$

As a result;

$$y[n] = \sum_{k=0}^{\infty} y_k[n - Lk].$$

3.6.2 Overlap-Save Method

Assume that the impulse response $h[n]$ has length P. The convolution of $x[n]$ and $h[n]$ using overlap-save method is achieved via the following steps.

(1) Pad the front of $x[n]$ by $P - 1$ zeros.
(2) Divide $x[n]$ into frames of length L such that the successor frame overlaps with the predecessor frame with $P - 1$ points.
(3) Let $x_k[n]$ be a frame, calculate the L point circular convolution of $x_k[n]$ and $h[n]$, i.e., calculate

$$y_k[n] = x_k[m](L)h[n].$$

(4) Discard the first $P - 1$ points of $y_k[n]$.
(5) Concatenate $y_k[n]$ and obtain $y[n]$, i.e., $y[n] = [y_0[n]y_1[n]\cdots]$.

Let's explain overlap-save method with an example.

Example 3.37 Using $h[n]$ and $x[n]$ given below, find the convolution of $h[n]$ and $x[n]$ using overlap-save method.

$$h[n] = \begin{bmatrix} -1 & 1 & 1 \end{bmatrix}$$

$$x[n] = \begin{bmatrix} 1 & 0 & -1 & 1 & 1 & -1 & 1 & 0 & 0 & 1 & 1 & -1 \end{bmatrix}$$

Take frame length as $L = 4$.

Solution 3.37 The length of the impulse response $h[n]$ is 3, i.e., $P = 3$. And frame length is $L = 4$ which is given the question, otherwise we can choose it according to our will.

Let's follow the steps of the overlap-save method for the calculation of convolution of $h[n]$ and $x[n]$.

(1) Add $P - 1 = 3 - 1 \rightarrow 2$ zeros to the beginning of $x[n]$. This is shown in

$$x[n] = [\; \underbrace{0\ \ 0}_{\substack{P-1\ \text{zeros}}}\ \ \ 1\ \ 0\ \ -1\ \ 1\ \ 1\ \ -1\ \ 1\ \ 0\ \ 0\ \ 1\ \ 1\ \ -1\;]$$

P − 1 zeros
are added
to the
beginning
of x[n]

(2) Divide $x[n]$ into frames such that frames overlap by $P - 1 = 2$ samples. This operation is illustrated in

$$\overbrace{}^{x_0[n]}\quad\overbrace{}^{x_2[n]}\quad\overbrace{}^{x_4[n]}\quad\overbrace{}^{x_6[n]}$$

$$x[n] = [0\ \ 0\ \ 1\ \ 0\ \ -1\ \ 1\ \ 1\ \ -1\ \ 1\ \ 0\ \ 0\ \ 1\ \ 1\ \ -1\ \ 0\ \ 0].$$

$$\underbrace{}_{x_1[n]}\quad\underbrace{}_{x_3[n]}\quad\underbrace{}_{x_5[n]}$$

where we padded the last divided frame by 2 zeros such that its length equals 4. The divided frames are separately written as

$$x_0[n] = [0\ \ 0\ \ 1\ \ 0]\quad x_1[n] = [1\ \ 0\ \ -1\ \ 1]\quad x_2[n] = [-1\ \ 1\ \ 1\ \ -1]$$
$$x_3[n] = [1\ \ -1\ \ 1\ \ 0]\quad x_4[n] = [1\ \ 0\ \ 0\ \ 1]\quad x_5[n] = [0\ \ 1\ \ 1\ \ -1]$$

(3) In step 3 we calculate the $L = 4$-point circular convolution of each frame with $h[n]$, i.e., we calculate

$$y_0[n] = h[n]\,(4)x_0[n]\quad y_1[n] = h[n]\,(4)x_1[n]\quad y_2[n] = h[n]\,(4)x_2[n]$$
$$y_3[n] = h[n]\,(4)x_3[n]\quad y_4[n] = h[n]\,(4)x_4[n]\quad y_5[n] = h[n]\,(4)x_5[n].$$

As a reminder we below provide the 4-points circular convolution of $x_0[n]$ and $h[n]$. N-point circular convolution of $x[n]$ and $h[n]$ is given as

$$y_0[n] = \sum_{k=0}^{N-1} x[k]h[(n-k)_N]. \tag{3.156}$$

For $N = 4$ when (3.156) is expanded, we obtain

$$y_0[n] = x[0]h[(n)_4] + x[1]h[(n-1)_4] + x[2]h[(n-2)_4] + x[3]h[(n-3)_4].$$

Since $N = 4$ we pad $h[n]$ by zeros such that its length equals $N = 4$ and $h[n]$ becomes as

$$h[n] = \begin{bmatrix} -1 & 1 & 1 & 0 \end{bmatrix}.$$

Noting that $h[(n - n_0)_4]$ is obtained rotating $h[n]$ to the right by n_0 units, we get the following expression for $y_0[n]$

$$y_0[n] = 0 \times h[(n)_4] + 0 \times h[(n-1)_4] + 1 \times h[(n-2)_4] + 0 \times h[(n-3)_4]$$

which leads to

$$y_0[n] = \begin{bmatrix} 1 & 0 & -1 & 1 \end{bmatrix}.$$

4-point circular convolution of each frame with $h[n]$ is given in (3.157).

$$y_0[n] = \begin{bmatrix} 1 & 0 & -1 & 1 \end{bmatrix} \quad y_1[n] = \begin{bmatrix} -1 & 2 & 2 & -2 \end{bmatrix} \quad y_2[n] = \begin{bmatrix} 1 & -3 & -1 & 3 \end{bmatrix}$$
$$y_3[n] = \begin{bmatrix} 0 & 2 & -1 & 0 \end{bmatrix} \quad y_4[n] = \begin{bmatrix} 0 & 2 & 1 & -1 \end{bmatrix} \quad y_5[n] = \begin{bmatrix} 0 & -2 & 0 & 3 \end{bmatrix}$$
$$y_6[n] = \begin{bmatrix} -1 & 2 & 0 & -1 \end{bmatrix}$$

$$(3.157)$$

(4) In step-4, we discard the first $P - 1 = 2$ samples from the beginning of each $y_k[n], k = 0, 1, 2, 3, 4$. This operation is illustrated in

$$y_0[n] = \begin{bmatrix} \underbrace{1 \quad 0}_{omit} & -1 & 1 \end{bmatrix} \rightarrow y_0[n] = \begin{bmatrix} -1 & 1 \end{bmatrix}$$

$$y_1[n] = \begin{bmatrix} \underbrace{-1 \quad 2}_{omit} & 2 & -2 \end{bmatrix} \rightarrow y_1[n] = \begin{bmatrix} 2 & -2 \end{bmatrix}$$

$$y_2[n] = \begin{bmatrix} \underbrace{1 \quad -3}_{omit} & -1 & 3 \end{bmatrix} \rightarrow y_2[n] = \begin{bmatrix} -1 & 3 \end{bmatrix}$$

$$y_3[n] = \begin{bmatrix} \underbrace{0 \quad 2}_{omit} & -1 & 0 \end{bmatrix} \rightarrow y_3[n] = \begin{bmatrix} -1 & 0 \end{bmatrix}$$

$$y_4[n] = \begin{bmatrix} \underbrace{0 \quad 2}_{omit} & -1 & 1 \end{bmatrix} \rightarrow y_4[n] = \begin{bmatrix} 1 & -1 \end{bmatrix}$$

$$y_5[n] = \begin{bmatrix} \underbrace{0 \quad -2}_{omit} & 0 & 3 \end{bmatrix} \rightarrow y_5[n] = \begin{bmatrix} 0 & 3 \end{bmatrix}$$

$$y_6[n] = \begin{bmatrix} \underbrace{-1 \quad 2}_{omit} & 0 & -1 \end{bmatrix} \rightarrow y_5[n] = \begin{bmatrix} 0 & -1 \end{bmatrix}.$$

(5) Finally in the last step, we concatenate the truncated sequences to find the convolution result, i.e.,

$$y[n] = [y_0[n]y_1[n]y_2[n]y_3[n]y_4[n]y_5[n]]$$

which leads to

$$y[n] = [-1 \quad 1 \quad 2 \quad -2 \quad -1 \quad 3 \quad -1 \quad 0 \quad 1 \quad -1 \quad 0 \quad 3 \quad 0 \quad -1].$$

Exercise: If $x[n] = [-1 \quad 1 \quad 1 \quad -1 \quad 1 \quad 1 \quad -1 -111-1-111-1-1]$ and

$h[n] = [1 \quad -1 \quad -1]$, calculate $x[n] * h[n]$

(a) Using overlap-add method.

(b) Using overlap-save method.

3.7 Computation of the Discrete Fourier Transform

3.7.1 Fast Fourier Transform (FFT) Algorithms

There are two types of Fast Fourier transform algorithm. These are:

(1) Decimation in time FFT algorithm.
(2) Decimation in frequency FFT algorithm.

Let's first explain decimation in time FFT algorithm then decimation in frequency FFT algorithm.

3.7.2 Decimation in Time FFT Algorithm

Before starting to the derivation of the algorithm, let's consider some motivating examples.

The DFT formula is

$$X[k] = \sum_{n=0}^{N-1} x[n]e^{-jk\frac{2\pi}{N}n}$$

where k takes values in the range $0, 1, \ldots, N - 1$, i.e., if $N = 4$, then the range of k is $0, 1, 2, 3$.

Example 3.38 If e_N^k is defined as $e_N^k = e^{-jk\frac{2\pi}{N}}$ $k \in Z$, write e_4^k for $k = 0, 1, 2, 3$ as a vector.

Solution 3.38 $e_4^k = \left[e^{-j0\frac{2\pi}{4}} \quad e^{-j1\frac{2\pi}{4}} \quad e^{-j2\frac{2\pi}{4}} \quad e^{-j3\frac{2\pi}{4}} \right]$ which can be simplified as

$$e_4^k = \begin{bmatrix} 1 & -j & -1 & j \end{bmatrix}$$

Exercise: Write e_8^k for $k = 0, 1, \ldots, 7$ as a vector.

Example 3.39 Given $x[n] = \begin{bmatrix} a & b \end{bmatrix}$ find 2-point DFT of $x[n]$.

Solution 3.39 Using the formula

$$X[k] = \sum_{n=0}^{N-1} x[n] e^{-jk\frac{2\pi}{N}n}, \quad k = 0, 1, \ldots, N-1$$

for $N = 2$, we get

$$X[k] = \sum_{n=0}^{1} x[n] e^{-jk\frac{2\pi}{2}n}, \quad k = 0, 1 \tag{3.158}$$

When (3.158) is expanded for $k = 0$ and $k = 1$, we get

$$X[0] = x[0] + x[1] \quad X[1] = x[0] + x[1]e^{j\pi} \rightarrow X[1] = x[0] - x[1].$$

Then 2-point DFT of $x[n] = \begin{bmatrix} a & b \end{bmatrix}$ is

$$X[k] = \begin{bmatrix} a+b & a-b \end{bmatrix}.$$

Example 3.40 If $x[n] = \begin{bmatrix} 3 & -2 \end{bmatrix}$, find 2-point DFT of $x[n]$.

Solution 3.40 Using $X[k] = \begin{bmatrix} a+b & a-b \end{bmatrix}$, we find the 2-point DFT of $x[n] = \begin{bmatrix} 3 & -2 \end{bmatrix}$ as

$$X[k] = \begin{bmatrix} 1 & 5 \end{bmatrix}.$$

Example 3.41 If $x[n] = \begin{bmatrix} a & b \end{bmatrix}$, find $X[k]$ for $k = 0, 1, 2, 3$.

Solution 3.41 Expanding the formula

$$X[k] = \sum_{n=0}^{N-1} x[n] e^{-jk\frac{2\pi}{N}n}$$

for $N = 2$ and $k = 0, 1, 2, 3$, we obtain

$$X[0] = \sum_{n=0}^{1} x[n] e^{-j0\frac{2\pi}{2}n}$$

$$X[1] = \sum_{n=0}^{1} x[n] e^{-j1\frac{2\pi}{2}n}$$

$$X[2] = \sum_{n=0}^{1} x[n] e^{-j2\frac{2\pi}{2}n}$$

$$X[3] = \sum_{n=0}^{1} x[n] e^{-j3\frac{2\pi}{2}n}.$$

If we look at the exponential terms in $X[0]$ and $X[2]$, we see that $e^{-j0\frac{2\pi}{2}n} = e^{-j2\frac{2\pi}{2}n}$ this means that

$$X[2] = X[0]$$

In a similar manner; we find that

$$X[3] = X[1]$$

Then $X[k]$ for $k = 0, 1, 2, 3$, happens to be

$$X[k] = \begin{bmatrix} X[0] & X[1] & \underbrace{X[2]}_{=X[0]} & \underbrace{X[3]}_{=X[1]} \end{bmatrix}$$

That is

$$X[k] = \begin{bmatrix} X[0] & X[1] & X[0] & X[1] \end{bmatrix}$$

And using our previous example results, we can write $X[k]$ as

$$X[k] = \begin{bmatrix} a+b & a-b & a+b & a-b \end{bmatrix}$$

Example 3.42 Calculate $X[k]$ for

$$x[n] = \begin{bmatrix} 1 & 3 & 2 & -1 \end{bmatrix}$$

using the DFT formula but take k range as $0, 1, \ldots, 7$ instead of $0, 1, \ldots, 3$.

Solution 3.42 Using the DFT formula the DFT coefficients for $k = 0, 1, \ldots, 7$ can be calculated as

$$X[0] = \sum_{n=0}^{3} x[n] e^{-j0\frac{2\pi}{4}n}$$

$$X[1] = \sum_{n=0}^{3} x[n] e^{-j1\frac{2\pi}{4}n}$$

$$X[2] = \sum_{n=0}^{3} x[n] e^{-j2\frac{2\pi}{4}n}$$

$$X[3] = \sum_{n=0}^{3} x[n] e^{-j3\frac{2\pi}{4}n}$$

$$X[4] = \sum_{n=0}^{3} x[n] e^{-j4\frac{2\pi}{4}n} \tag{3.159}$$

$$X[5] = \sum_{n=0}^{3} x[n] e^{-j5\frac{2\pi}{4}n}$$

$$X[6] = \sum_{n=0}^{3} x[n] e^{-j6\frac{2\pi}{4}n}$$

$$X[7] = \sum_{n=0}^{3} x[n] e^{-j7\frac{2\pi}{4}n}.$$

If we inspect the exponential terms in $X[0]$ and $X[4]$ in the equation set (3.159), we see that $e^{-j0\frac{2\pi}{4}n} = e^{-j4\frac{2\pi}{4}n}$ this means that

$$X[4] = X[0].$$

In a similar manner; we have

$$X[5] = X[1] \quad X[6] = X[2] \quad X[7] = X[3].$$

If we calculate $X[k]$ for $k = 0, 1, \ldots, 3$, we get

$$X[k] = [X[0] \quad X[1] \quad X[2] \quad X[3]].$$

And on the other hand if we calculate $X[k]$ for $k = 0, 1, \ldots, 7$, we get

$$X[k] = \left[X[0] \quad X[1] \quad X[2] \quad X[3] \quad \underbrace{X[4]}_{=X[0]} \quad \underbrace{X[5]}_{=X[1]} \quad \underbrace{X[6]}_{=X[2]} \quad \underbrace{X[7]}_{=X[3]} \right].$$

That is

$$X[k] = [X[0] \quad X[1] \quad X[2] \quad X[3] \quad X[0] \quad X[1] \quad X[2] \quad X[3]].$$

Using (3.159), $X[k]$ for $k = 0, 1, 2, 3$ can be calculated as

$$X[k] = [5 \quad -1 - j4 \quad 1 \quad -1 + j4]$$

and for $k = 0, 1, \ldots, 7$, it equals to

$$X[k] = [5 \quad -1 - j4 \quad 1 \quad -1 + j4 \quad 5 \quad -1 - j4 \quad 1 \quad -1 + j4].$$

In fact, the results of these examples are nothing but the main motivation for the derivation of the fast Fourier transform algorithm.

Now let's start the derivation of the fast Fourier transform algorithm.

Fast Fourier Transform Algorithm Derivation

We consider the DFT formula

$$X[k] = \sum_{n=0}^{N-1} x[n] e^{-jk\frac{2\pi}{N}n}. \tag{3.160}$$

Let's denote the exponential function $e^{-j\frac{2\pi}{N}}$ in (3.160) by e_N, i.e., $e_N = e^{-j\frac{2\pi}{N}}$, and the function e_N has the following properties.

(1) $e_N^2 = e_{N/2}$

This property comes from the definition directly, i.e.,

$$e_{N/2}^2 = e^{-j2\frac{2\pi}{N}}$$

which can be written as

$$e_{N/2}^2 = e^{-j\frac{2\pi}{N/2}} = e_{N/2} \rightarrow e_N^2 = e_{N/2}.$$

(2) $e_N^N = 1$ or more in general $e_N^{mN} = 1, m \in Z$

Again starting by the definition, we have

$$e_N = e^{-j\frac{2\pi}{N}} \rightarrow e_N^{mN} = e^{-jm\frac{2\pi N}{N}} \rightarrow e_N^{mN} = e^{-jm2\pi} \rightarrow e_N^{mN} = 1.$$

(3) $e_N^{(m+N)} = e_N^{(m)}$

Using property-2 we obtain

$$e_N^{(m+N)} = e_N^{(m)} \underbrace{e_N^{(N)}}_{=1} \rightarrow e_N^{(m+N)} = e_N^{(m)}.$$

This means that $f(m) = e_N^{(m)}$ is a periodic function, and its period equals to N, i.e., $f(m) = f(m+N)$.

Let's now derive the decimation in time FFT algorithm. We first write the DFT formula in terms of the defined function e_N as

$$X[k] = \sum_{n=0}^{N-1} x[n] e_N^{(kn)}, \quad k = 0, 1, \ldots, N-1 \tag{3.161}$$

which can be partitioned for even and odd n values as

$$X[k] = \sum_{n=0}^{N/2-1} x[2n] e_N^{(2kn)} + \sum_{n=0}^{N/2-1} x[2n+1] e_N^{(2n+1)k} \tag{3.162}$$

where the first term on the right side using the property $e_N^2 = e_{N/2}$ can be written as

$$\sum_{n=0}^{N/2-1} x[2n] e_N^{(2nk)} \rightarrow \sum_{n=0}^{N/2-1} x[2n] (e_N^2)^{nk} \rightarrow \sum_{n=0}^{N/2-1} x[2n] (e_{N/2})^{nk} \tag{3.163}$$

and the similarly the second term on the right side of (3.162) using the property $e_N^2 = e_{N/2}$ can be written as

$$\begin{aligned} \sum_{n=0}^{N/2-1} x[2n+1] e_N^{(2n+1)k} &\rightarrow \sum_{n=0}^{N/2-1} x[2n+1] e_N^{2nk} e_N^k \\ \rightarrow e_N^k \sum_{n=0}^{N/2-1} x[2n+1] e_N^{2nk} &\rightarrow e_N^k \sum_{n=0}^{N/2-1} x[2n+1] e_{N/2}^{nk} \end{aligned} \tag{3.164}$$

Then using the results (3.163) and (3.164), the DFT formula in (3.161) can be written as

$$X[k] = \underbrace{\sum_{n=0}^{N/2-1} x[2n] e_{N/2}^{nk}}_{G[k]} + e_N^k \underbrace{\sum_{n=0}^{N/2-1} x[2n+1] e_{N/2}^{nk}}_{H[k]} \quad k = 0, 1, \ldots, N-1$$

where the terms $G[k]$ and $H[k]$ are periodic with period $N/2$. Since $G[k]$ and $H[k]$ are calculated for $k = 0, 1, \ldots N - 1$ in $X[k]$ then $G[k]$ and $H[k]$ have repeated values for $k = 0, 1, \ldots N - 1$ as shown in

$$G[k] = \begin{bmatrix} \underbrace{g_0 \quad g_1 \quad g_2 \quad \cdots}_{\substack{\text{The first } N/2 \\ \text{samples}}} & \underbrace{g_0 \quad g_1 \quad g_2 \quad \cdots}_{\substack{\text{The second} N/2 \\ \text{samples}}} \end{bmatrix}$$

$$H[k] = \begin{bmatrix} \underbrace{h_0 \quad h_1 \quad h_2 \quad \cdots}_{\substack{\text{The first } N/2 \\ \text{samples}}} & \underbrace{h_0 \quad h_1 \quad h_2 \quad \cdots}_{\substack{\text{The second } N/2 \\ \text{samples}}} \end{bmatrix} .$$

And $G[k]$ $k = 0, 1, \ldots, N/2 - 1$ is the $N/2$ point DFT of the even numbered samples of $x[n]$, and $H[k]$ $k = 0, 1, \ldots, N/2 - 1$ is the $N/2$ point DFT of the odd numbered samples of $x[n]$.

Hence for the computation of $G[k]$ and $H[k]$ the k index range is first taken as $k = 0, 1, \ldots, N/2 - 1$. And $G[k]$ and $H[k]$ are calculated for $k = 0, 1, \ldots, N/2 - 1$. Let's denote the calculation results as

$$G[k] = \begin{bmatrix} g_0 & g_1 & \cdots & g_{N/2-1} \end{bmatrix} \quad H[k] = \begin{bmatrix} h_0 & h_1 & \cdots & h_{N/2-1} \end{bmatrix} \quad k = 0, 1, \ldots, N/2 - 1$$

Then $G[k]$ and $H[k]$ values for $k = 0, 1, \ldots, N - 1$ are obtained using

$$G[k] = \begin{bmatrix} g_0 & g_1 & \cdots & g_{N/2-1} & g_0 & g_1 & \cdots & g_{N/2-1} \end{bmatrix}$$
$$H[k] = \begin{bmatrix} h_0 & h_1 & \cdots & h_{N/2-1} & h_0 & h_1 & \cdots & h_{N/2-1} \end{bmatrix}$$

and they are combined in $X[k]$ via

$$X[k] = G[k] + w_N^k H[k] \quad k = 0, 1, \ldots, N - 1.$$

The partition performed for $X[k]$ can be done for $G[k]$ and $H[k]$ also. The calculation of $G[k]$ can be written as

$$G[k] = G_1[k] + w_N^k G_2[k] \quad k = 0, 1, \ldots, N/2 - 1$$

where $G_1[k]$ is the $N/4$ point DFT of the even numbered samples of $x[2n]$ and $G_2[k]$ is the $N/4$ point DFT of the odd numbered samples of $x[2n]$.

And the calculation of $H[k]$ can be written as

$$H[k] = H_1[k] + w_N^k H_2[k] \quad k = 0, 1, \ldots, N/2 - 1$$

where $H_1[k]$ is the $N/4$ point DFT of the even numbered samples of $x[2n + 1]$ and $H_2[k]$ is the $N/4$ point DFT of the odd numbered samples of $x[2n + 1]$.

This procedure can be carried out until we calculate 2-point DFT of the sequences obtained from $x[n]$.

Example 3.43 If $x[n] = \begin{bmatrix} a & b \end{bmatrix}$ find 2-point DFT of $x[n]$.

Solution 3.43 Using the formula

$$X[k] = \sum_{n=0}^{N-1} x[n]e^{-jk\frac{2\pi}{N}n}, \quad k = 0, 1, \ldots, N-1 \tag{3.165}$$

for $N = 2$, we get

$$X[k] = \sum_{n=0}^{1} x[n]e^{-jk\frac{2\pi}{2}n}, \quad k = 0, 1. \tag{3.166}$$

When (3.166) is expanded for $k = 0$ and $k = 1$, we obtain

$$X[0] = x[0] + x[1] \quad X[1] = x[0] + x[1]e^{j\pi} \rightarrow X[1] = x[0] - x[1]$$

which can be expressed in a more compact way as

$$X[k] = \begin{bmatrix} a+b & a-b \end{bmatrix}. \tag{3.167}$$

Example 3.44 If $x[n] = \begin{bmatrix} -1 & 4 \end{bmatrix}$, find 2-point DFT of $x[n]$.

Solution 3.44 $X[0] = -1 + 4 \rightarrow X[0] = 3 \quad X[1] = -1 - 4 \rightarrow X[1] = -5$.

Example 3.45 If $x[n] = \begin{bmatrix} 1 & 1 & -1 & 2 \end{bmatrix}$, find 4-point DFT of $x[n]$ using decimation in time FFT algorithm.

Solution-3.45: First we determine the even and odd numbered elements of $x[n]$ as in

$$x[n] = \begin{bmatrix} 1 & \overset{\uparrow}{1} & -1 & \overset{\uparrow}{2} \end{bmatrix}$$

where down-arrows indicate even numbered samples and up-arrows show odd numbered samples. And the even and odd numbered samples can be grouped into separate vectors as

$$x_e[n] = \begin{bmatrix} 1 & -1 \end{bmatrix} \quad x_o[n] = \begin{bmatrix} 1 & 2 \end{bmatrix}.$$

The 2-point DFT of $x_e[n]$ and $x_o[n]$ are calculated using DFT formula as

$$X_e[0] = 1 - 1 \rightarrow X_e[0] = 0 \quad X_e[1] = 1 - (-1) \rightarrow X_e[1] = 2$$
$$X_o[0] = 1 + 2 \rightarrow X_e[0] = 3 \quad X_o[1] = 1 - 2 \rightarrow X_e[1] = -1.$$

Hence, for $X_e[k]$ and $X_o[k]$ $k = 0, 1$; we have

$$X_e[k] = [0 \quad 2] \quad X_o[k] = [3 \quad -1]. \tag{3.168}$$

DFT of $x[n]$ can be written in terms of DFT of its even and odd samples as

$$X[k] = X_e[k] + w_N^k X_o[k] \quad k = 0, 1, \ldots, N - 1. \tag{3.169}$$

For $N = 4$ Eq. (3.169) is written as

$$X[k] = X_e[k] + w_4^k X_o[k] \quad k = 0, 1, \ldots, 4 - 1 \tag{3.170}$$

where

$$w_4^k = e^{-jk\frac{2\pi}{4}}.$$

And for $N = 4$ the vectors $X_e[k]$, $X_o[k]$ and w_4^k for $k = 0, 1, 2, 3$ can be calculated as

$$X_e[k] = [0 \quad 2 \quad 0 \quad 2] \quad X_0[k] = [3 \quad -1 \quad 3 \quad -1]$$
$$w_4^k = \left[e^{-j0\frac{2\pi}{4}} \quad e^{-j1\frac{2\pi}{4}} \quad e^{-j2\frac{2\pi}{4}} \quad e^{-j3\frac{2\pi}{4}} \right]. \tag{3.171}$$

And simplifying w_4^k, we get

$$w_4^k = [1 \quad -j \quad -1 \quad j].$$

Finally the vector $X[k]$ is obtained using (3.170) as in

$$X[k] = [0 \quad 2 \quad 0 \quad 2] + [- \quad -j \quad -1 \quad j] * [3 \quad -1 \quad 3 \quad -1]$$

where the vector product term

$$[1 \quad -j \quad -1 \quad j] * [3 \quad -1 \quad 3 \quad -1]$$

is calculated as

$$[1 \times 3 \quad (-j) \times (-1) \quad (-1) \times 3 \quad j \times (-1)].$$

Then $X[k]$ becomes as

$$X[k] = [0+1 \times 3 \quad 2+(-j) \times (-1) \quad 0+(-1) \times 3 \quad 2+j \times (-1)]$$

which has the final form

$$X[k] = [3 \quad 2+j \quad -3 \quad 2-j].$$

Example 3.46 If $x[n] = [1 \quad 1 \quad -1 \quad 2 \quad 1 \quad 3 \quad -1 \quad 2]$, find 8-point DFT of $x[n]$ using decimation in time FFT algorithm.

Solution 3.46 First, we divide the sequence $x[n]$ to its even and odd numbered elements as in

where down-arrows indicate even indexed samples and up-arrow shows odd indexed samples. And the even and odd indexed samples can be grouped into separate vectors as

$$x_e[n] = [1 \quad -1 \quad 1 \quad -1] \quad x_o[n] = [1 \quad 2 \quad 3 \quad 2].$$

Four-point DFT of $x_e[n]$ and $x_o[n]$ can be calculated as in the previous example as

$$X_e[k] = [0 \quad 0 \quad 4 \quad 0] \quad X_o[k] = [8 \quad -2 \quad 0 \quad -2] \quad k = 0, 1, \ldots, 4. \quad (3.172)$$

Then 8-point DFT of $x[n]$ is calculated through

$$X[k] = X_e[k] + w_8^k X_o[k] \quad k = 0, 1, \ldots, 7$$

where $w_8^k = e^{-jk\frac{2\pi}{8}}$. And the vectors $X_e[k]$, $X_o[k]$, w_8^k for $k = 0, 1, \ldots, 7$ with the help of (3.172) can be written as

$$X_e[k] = [0 \quad 0 \quad 4 \quad 0 \quad 0 \quad 0 \quad 4 \quad 0]$$

$$\begin{aligned} X_o[k] &= [8 \quad -2 \quad 0 \quad -2 \quad 8 \quad -2 \quad 0 \quad -2] \\ w_8^k &= [e^{-j0\frac{2\pi}{8}} \quad e^{-j1\frac{2\pi}{8}} \quad e^{-j2\frac{2\pi}{8}} \quad e^{-j3\frac{2\pi}{8}} \quad e^{-j4\frac{2\pi}{8}} \quad e^{-j5\frac{2\pi}{8}} \quad e^{-j6\frac{2\pi}{8}} \quad e^{-j7\frac{2\pi}{8}}] \end{aligned} \quad (3.173)$$

And combining the vectors in (3.173) using

$$X[k] = X_e[k] + w_8^k X_o[k] \quad k = 0, 1, \ldots, 7$$

we obtain the 8-point DFT of $x[n]$ as

$$X[k] = [8 \quad -1.4+j1.4 \quad 4 \quad 1.4+j1.4 \quad -8 \quad 1.4-j1.4 \quad 4 \quad 1.4-j1.4],$$
$$k = 0, 1, \ldots, 7.$$

Example 3.47 For the digital signal

$$x[n] = [1 \quad 1 \quad -1 \quad 2 \quad 1 \quad 3 \quad -1 \quad 2 \quad 1 \quad -1 \quad 2 \quad 1 \quad 3 \quad 0 \quad 1 \quad 2]$$

find 16-point DFT using decimation in time FFT algorithm.

Solution 3.47 First, we divide the signal to its even and odd indexed sequences as in

$$x_e[n] = [1 \quad -1 \quad 1 \quad -1 \quad 1 \quad 1 \quad 2 \quad 3 \quad 1]$$

$$x_o[n] = [1 \quad 2 \quad 3 \quad 2 \quad 1 \quad 1 \quad 0 \quad 2].$$

We can calculate 8-point DFT of $x_e[n]$ and $x_o[n]$ as in the previous example. Let the calculation results be denoted by $X_e[k]$ and $X_o[k]$, $k = 0, 1, \ldots, 7$. Then we can easily obtain $X_e[k]$ and $X_o[k]$ for $k = 0, 1, \ldots, 15$ by just repeating the elements obtained for $k = 0, 1, \ldots, 7$ and combine them using

$$X[k] = X_e[k] + w_{16}^k X_o[k] \quad k = 0, 1, \ldots, 15$$

where the exponential vector w_{16}^k, $k = 0, 1, \ldots, 15$ is calculated as

$$e_{16}^k = \left[e^{-j0\frac{2\pi}{16}} \quad e^{-j1\frac{2\pi}{16}} \quad e^{-j2\frac{2\pi}{16}} \quad e^{-j3\frac{2\pi}{16}} \quad e^{-j4\frac{2\pi}{16}} \quad e^{-j5\frac{2\pi}{16}} \quad e^{-j6\frac{2\pi}{16}} \quad e^{-j7\frac{2\pi}{16}} \quad e^{-j8\frac{2\pi}{16}} \quad e^{-j9\frac{2\pi}{16}} \right.$$
$$\left. e^{-j10\frac{2\pi}{16}} \quad e^{-j11\frac{2\pi}{16}} \quad e^{-j12\frac{2\pi}{16}} \quad e^{-j13\frac{2\pi}{16}} \quad e^{-j14\frac{2\pi}{16}} \quad e^{-j15\frac{2\pi}{16}} \right].$$

3.7.3 Decimation in Frequency FFT Algorithm

Before starting the derivation of decimation in frequency FFT algorithm let's solve some examples to become familiar with the terminology used in algorithm.

Example 3.48 If $x[n] = [1 \quad -2 \quad 3 \quad -6 \quad 4 \quad 2]$ $n = 0, 1, \ldots, 5$.

(a) Find $x[n]$ for $n = 0, 1, 2$.
(b) Find $x[n]$ for $n = 0, 1, \ldots, 4$.

Solution 3.48

(a) $x[n] = [1 \quad -2 \quad 3]$ $n = 0, 1, 2$

(b) $x[n] = [1 \quad -2 \quad 3 \quad -6 \quad 4] \quad n = 0, 1, \ldots, 4.$

Example 3.49 If $x[n] = [1 \quad -2 \quad 3 \quad -6 \quad 4 \quad 2] \quad n = 0, 1, \ldots, 5.$

(a) Find $x[n + N/2]$ for $n = 0, 1, 2$ and $N = 6.$

Solution 3.49 $x[n + N/2] = [-6 \quad 4 \quad 2] \quad n = 0, 1, 2$ and $N = 6$
 Note: $x[n]n = 0, 1, \ldots, N/2 - 1$ is the first half of the signal $x[n]$ and $x[n + N/2] \quad n = 0, 1, \ldots, N/2 - 1$ is the second half of the signal $x[n]$.

Example 3.50 If $x[n] = [1 \quad -2 \quad 3 \quad -6 \quad 4 \quad 2], \quad n = 0, 1, \ldots, 5.$

(a) Find $x[n] + x[n + N/2]$ for $n = 0, 1, 2$ and $N = 6.$
(b) Find $x[n] - x[n + N/2]$ for $n = 0, 1, 2$ and $N = 6.$

Solution 3.50 Using the results in previous example, we obtain

$$x[n] + x[n + N/2] = [-5 \quad 2 \quad 5] \quad x[n] - x\left[n + \frac{N}{2}\right] = [7 \quad -6 \quad 1].$$

Example 3.51 For $x[n] = [-2 \quad 1 \quad 3 \quad 5], \quad N = 4,$ find $x[n]e_N^n.$

Solution 3.51 Let's determine first e_N^n for $n = 0, 1, 2, 3.$ Using $e_N^n = e^{-jn\frac{2\pi}{N}}$ the vector form of e_N^n for $n = 0, 1, 2, 3$ can be written as

$$e_N^n = \left[e^{-j0\frac{2\pi}{4}} \quad e^{-j1\frac{2\pi}{4}} \quad e^{-j2\frac{2\pi}{4}} \quad e^{-j3\frac{2\pi}{4}}\right]$$

which can be simplified as

$$e_N^n = [1 \quad -j \quad -1 \quad -j].$$

Then the product signal $x[n]e_N^n$ for $n = 0, 1, 2, 3$ can be written as

$$x[n]e_N^n = [(-2) \times 1 \quad 1 \times (-j) \quad 3 \times (-1) \quad 5 \times j]$$

which yields

$$x[n]e_N^n = [-2 \quad -j \quad -3 \quad j5].$$

Example 3.52 $X[k] = [0 \quad 1 \quad 2 \quad 3 \quad 4 \quad 5 \quad 6 \quad 7]$ are the DFT coefficients of a digital signal $x[n]$. Write even and odd indexed samples of $X[k]$ as sequences.

Solution 3.52 Even indexed samples are

$$X[2k] = [0 \quad 2 \quad 4 \quad 6] \quad k = 0, 1, 2, 3$$

and odd indexed samples are

$$X[2k+1] = [1 \quad 3 \quad 5 \quad 7] \quad k = 0, 1, 2, 3.$$

Example 3.53 Even and odd indexed samples of the DFT coefficients of a digital signal are given as

$$X[2k] = [-1 \quad j \quad -2 \quad 3 \quad 1] \quad k = 0, 1, 2, 3, 4$$

$$X[2k+1] = [2 \quad 1+j \quad 2-j \quad 0 \quad -3] \quad k = 0, 1, 2, 3, 4$$

Find the DFT coefficient vector $X[k], k = 0, 1, \ldots, 9$.

Solution 3.53 Taking samples one by one from $X[2k]$ and $X[2k+1]$ in a sequential manner, we get the DFT coefficient vector

$$X[k] = [-1 \quad 2 \quad j \quad 1+j \quad -2 \quad 2-j \quad 3 \quad 0 \quad 1 \quad -3].$$

Let's now derive the decimation in frequency FFT algorithm.
Decimation in Frequency FFT Algorithm
In decimation in frequency FFT algorithm the even and odd indexed DFT coefficients are calculated separately. This operation is explained as follows. The DFT coefficients are calculated using

$$X[k] = \sum_{n=0}^{N-1} x[n] e_N^{kn} \quad k = 0, 1, \ldots, N-1 \qquad (3.174)$$

from which even indexed coefficients can be obtained via

$$X[2k] = \sum_{n=0}^{N-1} x[n] e_N^{2kn} \quad k = 0, 1, \ldots, N/2 - 1$$

where the summation term can be divided into two parts as

$$X[2k] = \sum_{n=0}^{N/2-1} x[n] e_N^{2kn} + \underbrace{\sum_{n=N/2}^{N-1} x[n] e_N^{2kn}}_{\sum_{n=0}^{\frac{N}{2}-1} x[n+\frac{N}{2}] e_N^{2k(n+\frac{N}{2})}} \quad k = 0, 1, \ldots, N/2 - 1. \qquad (3.175)$$

By changing the frontiers of the second summation expression in (3.175) we obtain

$$X[2k] = \sum_{n=0}^{N/2-1} x[n]e_N^{2kn} + \sum_{n=0}^{N/2-1} x\left[n+\frac{N}{2}\right]e_N^{2k(n+\frac{N}{2})} \quad k = 0, 1, \ldots, N/2 - 1 \quad (3.176)$$

where the exponential term $e_N^{2k(n+\frac{N}{2})}$ can be simplified as

$$e_N^{2k(n+\frac{N}{2})} = e_N^{2kn} \underbrace{e_N^{kN}}_{=1} \rightarrow e_N^{2k(n+\frac{N}{2})} = e_N^{2kn}$$

and making use of the $e_N^2 = e_{N/2}$ the expression for $X[2k]$ in (3.176) can be written as

$$X[2k] = \sum_{n=0}^{N/2-1} x[n]e_N^{2kn} + \sum_{n=0}^{N/2-1} x\left[n+\frac{N}{2}\right]e_{N/2}^{kn} \quad k = 0, 1, \ldots, N/2 - 1$$

which is further simplified as

$$X[2k] = \sum_{n=0}^{N/2-1} \left(x[n] + x\left[n+\frac{N}{2}\right]\right)e_{N/2}^{kn} \quad k = 0, 1, \ldots, N/2 - 1. \quad (3.177)$$

Equation (3.177) can be written in more compact form as

$$X[2k] = \sum_{n=0}^{N/2-1} x_1[n]e_{N/2}^{kn} \quad k = 0, 1, \ldots, N/2 - 1$$

where $x_1[n] = \left(x[n] + x\left[n+\frac{N}{2}\right]\right)$ $n = 0, 1, \ldots, N/2 - 1$.

In a similar manner the odd indexed coefficients of $X[k]$ can be obtained via

$$X[2k+1] = \sum_{n=0}^{N-1} x[n]e_N^{(2k+1)n} \quad k = 0, 1, \ldots, N/2 - 1$$

and proceeding as in the case of even indexed coefficients we obtain

$$X[2k+1] = \sum_{n=0}^{N/2-1} (x[n] - x[n+N/2])e_N^{(2k+1)n} \quad k = 0, 1, \ldots, N/2 - 1$$

which can also be written as

$$X[2k+1] = \sum_{n=0}^{N/2-1} (x[n] - x[n+N/2])e_N^n e_{N/2}^{kn} \quad k = 0, 1, \ldots, N/2 - 1$$

which can be written in more compact form as

$$X[2k+1] = \sum_{n=0}^{N/2-1} x_2[n]e_{N/2}^{kn} \quad k = 0, 1, \ldots, N/2 - 1$$

where $x_2[n] = (x[n] - x[n+N/2])e_N^n \quad n = 0, 1, \ldots, N/2 - 1$.
To sum it up;

$$X[2k] = \sum_{n=0}^{N/2-1} x_1[n]e_{N/2}^{kn} \quad k = 0, 1, \ldots, N/2 - 1$$

$$X[2k+1] = \sum_{n=0}^{N/2-1} x_2[n]e_{N/2}^{kn} \quad k = 0, 1, \ldots, N/2 - 1$$

where $x_1[n] = (x[n] + x[n+N/2]) \quad x_2[n] = (x[n] - x[n+N/2])e_N^n$
and $n = 0, 1, \ldots, N/2 - 1$, $e_N^n = e^{-j\frac{2\pi}{N}n}$.

Note: If the signal $x[n]$ is written as $x[n] = [A \quad B], n = 0, 1, \ldots, N - 1$ where A is the first half and B is the second half of $x[n]$, then

$$x[n] + x[n+N/2] = [A + B], \quad n = 0, 1, \ldots, N/2 - 1$$

and

$$x[n] - x[n+N/2] = [A - B], \quad n = 0, 1, \ldots, N/2 - 1$$

and

$$e_N^n \text{ for } n = 0, 1, \ldots, N/2 - 1$$

equals to

$$e_N^n = \left[e^{-j0\frac{2\pi}{N}} \quad e^{-j1\frac{2\pi}{N}} \quad \cdots \quad e^{-j\frac{N/2}{2}\frac{2\pi}{N}} \right].$$

Example 3.54 For $x[n] = [1 \quad 0 \quad 2 \quad -1]$ find DFT coefficients using decimation in frequency FFT method.

Solution 3.54 For the given sequence and $N = 4$ and let's first find the signals $x_1[n]$ and $x_2[n]$ given as

$$x_1[n] = (x[n] + x[n+N/2]) \quad x_2[n] = (x[n] - x[n+N/2])e_N^n$$

$$n = 0, 1, \ldots, N/2 - 1.$$

The signal $x_1[n]$ is obtained by summing the first and second half parts of $x[n]$ as follows

$$x[n] = \underbrace{\begin{bmatrix} 1 & 0 \end{bmatrix}}_{\substack{First \\ Half}} \ \underbrace{\begin{bmatrix} 2 & -1 \end{bmatrix}}_{\substack{Second \\ Half}}$$

$$x_1[n] = \begin{bmatrix} 1 & 0 \end{bmatrix} + \begin{bmatrix} 2 & -1 \end{bmatrix} \rightarrow x_1[n] = \begin{bmatrix} 1+2 & 0-1 \end{bmatrix}.$$

To calculate $x_2[n]$, we first compute e_N^n for $N = 4$ and $n = 0, 1$ as in

$$e_4^n = \left[e^{-j0\frac{2\pi}{4}} e^{-j1\frac{2\pi}{4}} \right] \rightarrow e_4^n = \begin{bmatrix} 1 & -j \end{bmatrix}.$$

And $x[n] - x[n+N/2]$ for $N = 4$ is calculated by subtracting the first and second half parts of $x[n]$ as follows

$$x[n] - x[n+2] = \begin{bmatrix} 1 & 0 \end{bmatrix} - \begin{bmatrix} 2 & -1 \end{bmatrix} \rightarrow x[n] - x[n+2] = \begin{bmatrix} -1 & 1 \end{bmatrix}.$$

Thus $x_2[n]$ is calculated as

$$x_2[n] = (x[n] - x[n+2])e_4^n \rightarrow x_2[n] = \begin{bmatrix} -1 & 1 \end{bmatrix} * \begin{bmatrix} 1 & -j \end{bmatrix}$$

which yields

$$x_2[n] = \begin{bmatrix} -1 & -j \end{bmatrix}.$$

Next, we calculate the DFT coefficients of $x_1[n]$ and $x_2[n]$ as follows

$$x_1[n] = \begin{bmatrix} 3 & -1 \end{bmatrix} \rightarrow X_1[k] = \begin{bmatrix} 3 & -1 & 3 & +1 \end{bmatrix}$$

$$x_2[n] = \begin{bmatrix} 1 & -j \end{bmatrix} \rightarrow X_2[k] = \begin{bmatrix} -1 & -j & -1 & +j \end{bmatrix}$$

where $X_1[k]$ and $X_2[k]$ for $k = 0, 1$ corresponds to $X[2k+1]$ and $X[2k]$ respectively. Then we get

$$X[2k+1] = \begin{bmatrix} \underbrace{2}_{X[1]} & \underbrace{4}_{X[3]} \end{bmatrix}$$

$$X[2k] = \begin{bmatrix} \underbrace{-1-j}_{X[0]} & \underbrace{-1+j}_{X[2]} \end{bmatrix}$$

As a result $X[k]$ becomes as

$$X[k] = [-1-j \quad 2 \quad -1+j \quad 4].$$

Now let's generalize this example employing parameters instead of using the numeric values.

Example 3.55 For $x[n] = [a \quad b \quad c \quad d]$, find DFT coefficients using decimation in frequency FFT method.

Solution 3.55 For the given sequence, let's first find the signals $x_1[n]$ and $x_2[n]$ given as

$$x_1[n] = (x[n] + x[n+N/2]) \quad x_2[n] = (x[n] - x[n+N/2])e_N^n$$

$$n = 0, 1, \ldots, N/2 - 1.$$

The signal $x_1[n]$ is obtained by summing the first and second half parts of $x[n]$ as follows

$$x[n] = \underbrace{[a \quad b}_{\substack{First \\ Half}} \quad \underbrace{c \quad d}_{\substack{Second \\ Half}} \,]$$

$$x_1[n] = [a \quad b] + [c \quad d] \rightarrow x_1[n] = [a+c \quad b+d].$$

To calculate $x_2[n]$, we first compute e_N^n for $N = 4$ and $n = 0, 1$ as follows

$$e_4^n = \left[e^{-j0\frac{2\pi}{4}} \quad e^{-j1\frac{2\pi}{4}} \right] \rightarrow e_4^n = [1 \quad -j].$$

And $x[n] - x[n+N/2]$ for $N = 4$ is calculated by subtracting the first and second half parts of $x[n]$ as in

$$x[n] - x[n+2] = [a \quad d] - [c \quad d] \rightarrow x[n] - x[n+2] = [a-c \quad b-d].$$

Thus $x_2[n]$ can be calculated as

$$x_2[n] = (x[n] - x[n+2])e_4^n \rightarrow x_2[n] = [-1 \quad 1] * [1 \quad -j]$$

which yields

$$x_2[n] = [a-c \quad -j(b-d)].$$

Next, we calculate the DFT coefficients of $x_1[n]$ and $x_2[n]$, i.e.,

$$x_1[n] = [a+c \quad b+d] \rightarrow X_1[k] = [a+c+b+d \quad a+c-b-d],$$
$$x_2[n] = [a-c \quad -j(b-d)] \rightarrow X_2[k] = [a-c-jb+jd \quad a-c+jb-jd]$$

where $X_1[k]$ and $X_2[k]$ for $k = 0, 1$ corresponds to $X[2k+1]$ and $X[2k]$ respectively. That is

$$X[2k+1] = \left[\underbrace{a+c+b+d}_{X[1]} \quad \underbrace{a+c-b-d}_{X[3]}\right],$$

$$X[2k] = \left[\underbrace{a-c-jb+jd}_{X[0]} \quad \underbrace{a-c+jb-jd}_{X[2]}\right].$$

As a result $X[k]$ becomes as

$$X[k] = [a-c-jb+jd \quad a+c+b+d \quad a-c+jb-jd \quad a+c-b-d].$$

Example 3.56 For $x[n] = [2 \quad 1 \quad 1 \quad -1 \quad 3 \quad 0 \quad 1 \quad -2]$, find 8-point DFT coefficients using decimation in frequency FFT method.

Solution 3.56 The first and second half parts of $x[n]$ are shown in

$$x[n] = \left[\underbrace{2 \quad 1 \quad 1 \quad -1}_{First\ Half} \quad \underbrace{3 \quad 0 \quad 1 \quad -2}_{Second\ Half}\right].$$

The signals

$$x[n]+x[n+N/2] \quad x[n]+x[n+N/2] \quad e_N^n$$

for $N = 8$, $n = 0, 1, \ldots, 3$ can be calculated as

$$x[n]+x[n+4] = [2 \quad 1 \quad 1 \quad -1] + [3 \quad 0 \quad 1 \quad -2] \rightarrow [4 \quad 1 \quad 2 \quad -3]$$

$$x[n]-x[n+4] = [2 \quad 1 \quad 1 \quad -1] - [3 \quad 0 \quad 1 \quad -2] \rightarrow [-1 \quad 1 \quad 0 \quad 1]$$

$$e_8^n = \left[e^{-j0\frac{2\pi}{8}} \quad e^{-j1\frac{2\pi}{8}} \quad e^{-j2\frac{2\pi}{8}} \quad e^{-j3\frac{2\pi}{8}}\right] \rightarrow e_8^n = \left[1 \quad e^{-j\frac{\pi}{4}} \quad e^{-j\frac{\pi}{2}} \quad e^{-j\frac{3\pi}{4}}\right]. \quad (3.178)$$

Using the results in (3.178), we can obtain the signals $x_1[n]$ and $x_2[n]$ as in

$$x_1[n] = [4 \quad 1 \quad 2 \quad -3]$$
$$x_2[n] = (x[n]-x[n+4])e_8^n$$

$$x_2[n] = [-1 \quad 1 \quad 0 \quad 1] * \left[1 \quad e^{-j\frac{\pi}{4}} \quad e^{-j\frac{\pi}{2}} \quad e^{-j\frac{3\pi}{4}}\right] \rightarrow$$

Hence we obtained the signals

$$x_1[n] = [4 \quad 1 \quad 2 \quad -3]$$

$$x_2[n] = \left[-1 \quad e^{-j\frac{\pi}{4}} \quad 0 \quad e^{-j\frac{3\pi}{4}}\right].$$

The DFT coefficients of $x_1[n]$ and $x_2[n]$ can be found using the decimation in frequency FFT algorithm as in the previous example. Let's denote the DFT coefficients of $x_1[n]$ and $x_2[n]$ as $X_1[k]$ and $X_2[k]$ which can be found as

$$X_1[k] = [4 \quad 2 - j4 \quad 8 \quad 2 + j4]$$

$$X_2[k] = [-1 - j2.8 \quad 1 - j2.8 \quad 1 + j2.82 \quad -1 + j2.82].$$

The Fourier coefficients of $x[n]$, i.e., $X[k]$ are related to $X_1[k]$ and $X_2[k]$ via

$$X[2k+1] = X_1[k] \quad X[2k] = X_2[k].$$

Then we have

$$X[2k+1] = [4 \quad 2 - j4 \quad 8 \quad 2 + j4]$$
$$X[2k] = [-1 - j2.8 \quad 1 - j2.8 \quad 1 + j2.82 \quad -1 + j2.82]$$

and $X[k]$ becomes as

$$X[k] = [-1 - j2.8 \quad 4 \quad 1 - j2.8 \quad 2 - j4 \quad 1 + j2.8 \quad 8 \quad -1 + j2.82 \quad 2 + j4].$$

3.8 Total Computation Amount of the FFT Algorithm

Consider the calculation of the following expression

$$x^2 + xy.$$

Now we ask the question: How many mathematical operations are needed for the calculation of $x^2 + xy$?

The answer is as follows.

For the computation of x^2, one multiplicative operation is needed.

For the computation of xy, one multiplicative operation is needed.

For the computation of $x^2 + xy$, two multiplicative operations and one additive operation is needed.

Hence, for the computation of $x^2 + xy$, three mathematical operations are needed.

Now consider the equality

$$x^2 + xy = x(x+y).$$

And we ask the same question: How many mathematical operations are needed for the calculation of $x(x+y)$?

It is obvious that for the calculation of $x(x+y)$, one additive operation and one multiplicative operation is needed. And the total number of mathematical operations for the calculation of $x(x+y)$ equals to two.

As a result; for $x^2 + xy$, three mathematical operations are needed, and for $x(x+y)$, two mathematical operations are needed. The latter one is preferable since it involves less computation amount.

Decimation in time and decimation in frequency FFT algorithms are invented to decrease the computation amount for the calculation of discrete transform coefficients $X[k]$ of a digital signal $x[n]$.

We can express the total computation saving for the calculation of DFT coefficients $X[k]$ of a digital signal $x[n]$ when FFT algorithms are employed other than the direct calculation approach. For illustration purposes, in the next section, we will first calculate the total computation amount for the evaluation of DFT coefficients $X[k]$ of a digital sequence $x[n]$.

Total Computation Amount of the Direct DFT Calculation:

Let's start the discussion with an example.

Example 3.57 For $N = 3$, find the total computation amount of the DFT formula

$$X[k] = \sum_{n=0}^{N-1} x[n] e^{-jk\frac{2\pi}{N}n}, \quad k = 0, 1, \ldots, N-1.$$

Solution 3.57 For $N = 3$ the DFT formula takes the form

$$X[k] = \sum_{n=0}^{2} x[n] e^{-jk\frac{2\pi}{N}n}, \quad k = 0, 1, 2$$

which is expanded as

$$X[0] = \sum_{n=0}^{2} x[n] e^{-j0\frac{2\pi}{3}n}$$

$$X[1] = \sum_{n=0}^{2} x[n] e^{-j1\frac{2\pi}{3}n} \tag{3.179}$$

$$X[2] = \sum_{n=0}^{2} x[n] e^{-j2\frac{2\pi}{3}n}.$$

When the summation terms in (3.179) are expanded, we get

$$X[0] = x[0]e^{-j0\frac{2\pi}{3}0} + x[1]e^{-j1\frac{2\pi}{3}0} + x[2]e^{-j2\frac{2\pi}{3}0}$$
$$X[1] = x[0]e^{-j\frac{2\pi}{3}0} + x[1]e^{-j\frac{2\pi}{3}1} + x[2]e^{-j\frac{2\pi}{3}2} \qquad (3.180)$$
$$X[2] = x[0]e^{-j2\frac{2\pi}{3}0} + x[1]e^{-j2\frac{2\pi}{3}1} + x[2]e^{-j2\frac{2\pi}{3}2}.$$

As can be seen from (3.180) for the calculation of each coefficient in (3.180), three multiplicative and two additive operations are required. Then the total number of multiplicative operations for the calculation of all the coefficients is $3 \times 3 = 9$ and the total number of additive operations for the calculation of all the coefficients is $3 \times 2 = 6$.

In general, for the calculation of N-point DFT $X[k]$ coefficients of a digital signal $x[n]$, N^2 multiplicative operations and $N \times (N-1)$ additive operations are needed. The total computation amount is

$$N^2 + N \times (N-1) \cong 2N^2.$$

Now let's consider the total computation amount of the decimation in time FFT algorithm.

Total Computation Amount of the Decimation in Time FFT Algorithm
Let's solve some examples to get familiar with the expressions appearing in this section.

Example 3.58 Let $N = 2^4$, we will divide N by 2 and divide the division result by 2 also and repeat this procedure until the result equals 2. How many divisions need to be performed?

Solution 3.58 $2^4/2 = 2^3 \rightarrow 2^3/2 = 2^2 \rightarrow 2^2/2 = 2$

As it is clear from the above result, 3 division operations are needed.

Note: If $N = 2^v$, then v division operations are needed to get 2 at the end of successive divisions.

Example 3.59 $a \leftarrow b$ means that whenever you see the a term replace it by b term in a mathematical expression. Let's define

$$N^2 \leftarrow N + 2\left(\frac{N}{2}\right)^2 \quad \text{if } N > 2$$
$$N^2 \leftarrow N \qquad\qquad\; \text{if } N = 2. \qquad (3.181)$$

Using (3.181), calculate the term that should be replaced for 8^2.

Solution 3.59 Using the definition we get

$$8^2 \leftarrow 8 + 2(4)^2 \quad (4)^2 \leftarrow 4 + 2(2)^2 \quad (2)^2 \leftarrow 2.$$

And for the expression $8^2 \leftarrow 8 + 2(4)^2$ inserting $4 + 2(2)^2$ for $(4)^2$, we get

$$8^2 \leftarrow 8 + 2(4 + 2(2)^2)$$

where replacing $(2)^2$ by 2, we obtain

$$8^2 \leftarrow 8 + 2(4 + 2 \times 2)$$

which is simplified as

$$8^2 \leftarrow 24. \tag{3.182}$$

In (3.182) the obtained result equals to $8 \times \log_2 8$.

Note: In general;

$$N^2 \leftarrow \underbrace{N + 2\left(\frac{N}{2}\right)^2}_{=N \times \log_2 N}.$$

Now let's consider the computation amount for the decimation in time FFT algorithm.

In decimation in time FFT algorithm DFT coefficients $X[k]$ of $x[n]$ are calculated using

$$X[k] = G[k] + w_N^k H[k] \quad k = 0, 1, \dots, N-1 \tag{3.183}$$

where $G[k]$ and $H[k]$ are the $N/2$ point DFT coefficients of even and odd indexed samples of $x[n]$. The calculation complexities for the terms appearing on the right hand side of (3.183) can be states as:

$$G[k] \rightarrow \left(\frac{N}{2}\right)^2 \text{multiplicative and} \left(\frac{N}{2}\right)\left(\frac{N}{2} - 1\right) \text{additive operations.}$$

$$H[k] \rightarrow \left(\frac{N}{2}\right)^2 \text{multiplicative and} \left(\frac{N}{2}\right)\left(\frac{N}{2} - 1\right) \text{additive operations.}$$

$$w_N^k H[k] \rightarrow N \text{multiplicative operations.}$$

And lastly for the summation of $G[k]$ and $w_N^k H[k]$ terms in (3.183), we need N more additive operations.

Thus; the total number of multiplicative operations is

$$\left(\frac{N}{2}\right)^2 + \left(\frac{N}{2}\right)^2 + N = N + \frac{N^2}{2}$$

which is less then N^2, i.e.,

$$\frac{N^2}{2} + N < N^2$$

and the total number of additive operations is

$$\left(\frac{N}{2}\right)\left(\underbrace{\frac{N}{2} - 1}_{ignore}\right) + \left(\frac{N}{2}\right)\left(\underbrace{\frac{N}{2} - 1}_{ignore}\right) + N \approx N + \frac{N^2}{2}$$

which is less then N^2.

Hence considering the total number of multiplicative and additive operations the computational complexity is less in decimation in time FFT algorithm.

Now let's consider the number of multiplicative operations

$$N + \frac{N^2}{2}$$

which can be written as

$$N + 2\left(\frac{N}{2}\right)^2 \tag{3.184}$$

which is replaced for N^2 when decimation in time FFT algorithm is applied.

The term $\left(\frac{N}{2}\right)$ in the (3.184) indicates the FFT computational complexity of $G[k]$ and $H[k]$. If decimation in time algorithm is applied for the calculation of $G[k]$ and $H[k]$, we can replace $\left(\frac{N}{2}\right)^2$ in (3.184) by

$$\frac{N}{2} + 2\left(\frac{N}{4}\right)^2$$

yielding

$$N + 2\left(\frac{N}{2} + 2\left(\frac{N}{4}\right)^2\right) = N + N + 4\left(\frac{N}{4}\right)^2$$

and proceeding in a similar manner and replacing $\left(\frac{N}{4}\right)^2$ by $\frac{N}{4} + 2\left(\frac{N}{8}\right)^2$, we get

$$N + N + 4\left(\frac{N}{4} + 2\left(\frac{N}{8}\right)^2\right) = N + N + N + 8\left(\frac{N}{8}\right)^2.$$

This procedure is carried out until we reach to 2-point FFT calculation. If $N = 2^v$, i.e., $v = \log_2 N$ the successive division process results in

$$\underbrace{N+N+\cdots+N}_{v\ terms}=vN$$

where replacing v by $\log_2 N$, we get

$$N\log_2 N$$

as the number of multiplicative operations required for the calculation of DFT coefficients of $x[n]$ using decimation in time FFT algorithm.

A similar procedure can be carried out to find the total number of additive operations required for the calculation of DFT coefficients of $x[n]$ using decimation in time FFT algorithm.

3.9 Problems

(1) If

$$x[n] = [1.0\quad 1.6\quad 2\quad 2.32\quad 2.58\quad 2.84\quad 3\quad 3.16\quad 3.4\quad 3.44\quad 3.58\quad 3.74\quad 3.84\quad 3.90],$$

then find $x[n-2], x[n+3], x[-n-2], x[2n-2], x[-2n-2],$ $x\left[\frac{n}{2}-2\right], x\left[-\frac{n}{3}-2\right].$

(2) One period of the periodic signal $\tilde{x}[n]$ around origin is $x[n] = [-1\quad 2\quad 1\quad -1\quad 2]$. Find one period of $\tilde{x}[n-2], \tilde{x}[n+2], \tilde{x}[-n],$ $\tilde{x}[-n-2], \tilde{x}[2n], \tilde{x}[-2n], \tilde{x}[2n+2], \tilde{x}[-2n+3].$

(3) One period of the periodic signal $\tilde{x}[n]$ around origin is $x[n] = [-1\quad 2\quad 1\quad -1]$. Find $\tilde{x}[n] * \tilde{x}[n].$

(4) If $x[n] = [-1\quad 2\quad 1\quad -1\quad 1]$, find $x[(n)_5], x[(-n)_5], x[(1-n)_5],$ $x[(3-n)_5], x[(n+2)_5], x[(-n+2)_5], x[(2n)_5], x[(-n-3)_5],$ for $0 \le n \le 4.$

(5) If $x[n] = [-1\quad 2\quad 1\quad -1\quad 1]$, find $x[(n)_3], x[(1-n)_3], x[(n+2)_3],$ $x[(-n+2)_3], x[(2n)_3], x[(-n-3)_3],$ for $0 \le n \le 2.$

Fig. 3.34 One period of the Fourier transform of the aperiodic signal $x[n]$

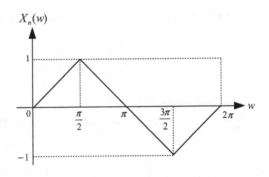

(6) Calculate 4-point DFT of $x[n] = [2 \quad -3 \quad 3 \quad 4]$.

(7) Calculate 6-point DFT of $x = [2 \quad -3 \quad 3 \quad 4]$.

(8) Find 5-point circular convolution of $x[n] = [1 \quad -1 \quad 2 \quad -1]$ and $y[n] = [1 \quad 0 \quad 3]$.

(9) If $x[n] = \begin{bmatrix} -1 & 0 & \underset{n=0}{1} & -1 & 2 & 1 \end{bmatrix}$, find

$$x[n] \; \textcircled{7} \; x[n] =?$$

(10) If $x[n] = [1 \quad 2 \quad 0 \quad -3 \quad -1]$ and

$$\tilde{x}[n] = \sum_{k=-\infty}^{\infty} x[n - 5k],$$

draw one period of the following signals.

(a) $\tilde{x}[n]$ (b) $\tilde{x}[2 - n]$ (c) $\tilde{x}[n - 2]$ (d) $\tilde{x}[2n - 1]$.

(11) One period of the Fourier transform of the aperiodic signal $x[n]$ is shown in Fig. 3.34.

(a) Find 8-point DFT of $x[n]$ i.e., $X[k] = ?$

(b) Using the DFT coefficients calculated in part (a), find $x[n]$ employing inverse DFT formula.

(12) Find the convolution of $x[n] = [1 \quad 0 \quad 1 \quad 1 \quad -1 \quad 0 \quad 1 \quad 2 \quad 3 \quad 1 \quad 1{-}1 \quad 4 \quad 1 \quad 2 \quad -1]$ and $h[n] = [1 \quad -1 \quad 1]]$ using overlap-add and overlap-save methods.

(13) Find the DFT of $x[n] = [1 \quad 0 \quad 1 \quad 1 \quad -1 \quad 0 \quad 1 \quad 2]$ using decimation in time FFT algorithm.

(14) Find the DFT of $x[n] = [1 \quad 0 \quad 1 \quad 2 \quad -1 \quad 0 \quad 1 \quad 2]$ using decimation in frequency FFT algorithm.

Chapter 4
Analog and Digital Filter Design

In this chapter, we will study analog and digital filter design techniques. A filter is nothing but a linear time invariant (LTI) system. Any LTI system can be described using its impulse response. If the impulse response of a LTI system is known, then for any arbitrary input the system output can be calculated by taking the convolution of the impulse response and arbitrary input. This also means that filtering operation is nothing but a convolution operation. And filter design is nothing but finding the impulse response of a linear time invariant system. For this purpose, we can work either in time domain or frequency domain.

Filter systems are designed to block some input frequencies and pass others. For this reason, filter design studies are usually done in frequency domain. Fourier transform of the impulse response of the filter system is called the transfer function of the filter. To find the transfer function of filters, a number of techniques are proposed in the literature. In this chapter, we will study the most widely known techniques in the literature.

Filters are divided into two main categories. These are analog filters and digital filters. In science world, more studies on analog filter design techniques are available considering the digital filter design methods. For this reason, so as to design a digital filter, usually digital filter specifications are transferred to analog domain, and analog filter design is performed then the designed analog filter is transferred to digital domain.

4.1 Review of Systems

In this chapter, we will study analog and digital filter design. Before studying filter design techniques, we will first review some fundamental concepts. We will follow the following outline in this chapter.

© Springer Nature Singapore Pte Ltd. 2018
O. Gazi, *Understanding Digital Signal Processing*, Springer Topics in Signal Processing 13, DOI 10.1007/978-981-10-4962-0_4

Fig. 4.1 A digital system

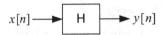

$x[n] \longrightarrow \boxed{H} \longrightarrow y[n]$

(a) Review of Systems.
(b) Review of Z-Transform.
(c) Review of Laplace Transform.
(d) Transformation between Continuous and Discrete Systems.
(e) Analogue Filter Design.
(f) IIR Digital Filter Design.
(g) FIR Digital Filter Design.

Hence, as outlined above before studying analog filter design, we will review some fundamental concepts, such as linear systems, z-transform, Laplace transform, and transformation between continuous and discrete systems.

The system given in the Fig. 4.1 has input $x[n]$ and output $y[n]$. And the relation between input and output can be indicated as $y[n] = H\{x[n]\}$.

Linearity:

The system H is a linear system if for the linear combination of the inputs the system output equals to the linear combination of the individual output. This is graphically illustrated in Fig. 4.2.

Mathematically the linearity property for the system H is expressed as

$$H\{ax_1[n] + bx_2[n]\} = aH\{x_1[n]\} + bH\{x_2[n]\}. \tag{4.1}$$

Time Invariance:

The system H is time invariant if

$$y[n - n_0] = H\{x[n - n_0]\} \tag{4.2}$$

Linear and Time Invariant System:

If a system is both linear and time invariant, then the system is called linear time invariant system, i.e., LTI system.

For a linear time invariant system denoted by H, the impulse response is defined as

$$h[n] = H\{\delta[n]\} \tag{4.3}$$

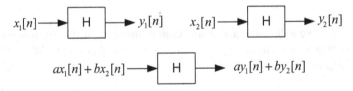

Fig. 4.2 Linear system

Fig. 4.3 Impulse response
and output of a linear time
invariant system

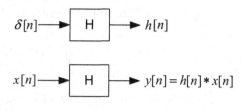

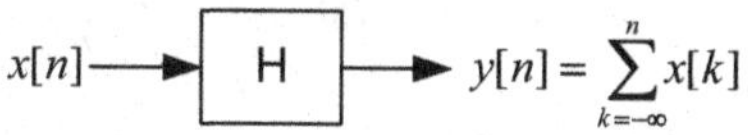

Fig. 4.4 A LTI system

and the output of a LTI system for an arbitrary input is defined as

$$y[n] = h[n] * x[n] \tag{4.4}$$

where $*$ denotes the convolution operation and it is evaluated as

$$h[n] * x[n] = \sum_{k=-\infty}^{\infty} h[k]x[n-k]. \tag{4.5}$$

This property graphically illustrated as in the following Fig. 4.3
Causality:
The signal $x[n]$ is causal if $x[n] = 0$ for $n < 0$.
The linear time invariant system denoted by H is causal if $h[n] = 0$ for $n < 0$.
Difference Equations for LTI Systems:
The relationship between the input and the output of a LTI system can be represented by difference equations as in

$$\sum_{k=0}^{N} a[k]y[n-k] = \sum_{k=0}^{M} b[k]x[n-k] \tag{4.6}$$

where $y[n]$ is the system output and $x[n]$ is the system input.

Example 4.1 The system H given in Fig. 4.4 is a LTI system.

(a) Write a difference equation between system input and output.
(b) Determine whether the system is causal or not.

Solution 4.1

(a) The relation between system input $x[n]$ and system output $y[n]$ is given as

$$y[n] = \sum_{k=-\infty}^{n} x[k]. \tag{4.7}$$

Using (4.7) then the shifted signal $y[n-1]$ can be calculated as

$$y[n-1] = \sum_{k=-\infty}^{n-1} x[k]. \tag{4.8}$$

Taking the difference of $y[n]$ in (4.7) and $y[n-1]$ in (4.8), we get

$$y[n] - y[n-1] = x[n]. \tag{4.9}$$

Using (4.7) the impulse response of the system can be calculated as

$$h[n] = \sum_{k=-\infty}^{n} \delta[k].$$
$$= u[n] \tag{4.10}$$

where it is seen that $h[n] = 0$ for $n<0$, which means that H is a causal system.

4.1.1 Z-Transform

For a digital sequence $x[n]$ the Z-transform is defined as

$$X(z) = \sum_{n=-\infty}^{\infty} x[n]z^{-n} \tag{4.11}$$

where the complex numbers $z = re^{jw}$ are chosen from a circle of radius r in complex plane. Substituting $z = re^{jw}$ into (4.11), we obtain

$$X\left(re^{jw}\right) = \sum_{n=-\infty}^{\infty} \left(x[n]r^{-n}\right)e^{-jwn} \tag{4.12}$$

which converges to a finite summation if

$$\sum_{n=-\infty}^{\infty} |x[n]r^{-n}| < \infty. \tag{4.13}$$

Since $z = re^{jw}$ then $|z| = r$ and according to (4.13) we see that the Z-transform converges only for a set of z-values and this set of z-values constitute a region in the complex plane. And this region is called region of convergence for $X(z)$.

The Properties of the Region of Convergence:

If $X(z) = \frac{P(z)}{Q(z)}$, then the roots of $P(z) = 0$ are called the zeros of $X(z)$ and the roots of $Q(z) = 0$ are called the poles of $X(z)$. The region of convergence of $X(z)$ has the following properties.

(1) The ROC does not contain any poles.
(2) Fourier transform of $x[n]$ exists if the ROC of $X(z)$ covers the unit circle.
(3) For a right sided sequence, the ROC extends outward from the outermost pole of $X(z)$.
(4) For a left sided sequence, the ROC extends inward from the innermost finite pole of $X(z)$.
(5) For a finite sequence, the ROC is a ring.

Example 4.2 For $x[n] = -\alpha^n u[-n-1]$, find $X(z)$.

Solution 4.2 Using the definition $X(z) = \sum_{n=-\infty}^{\infty} x[n]z^{-n}$ for the given signal, we obtain

$$X(z) = \sum_{n=-\infty}^{\infty} -\alpha^n u[-n-1]z^{-n}$$

where $u[-n-1]$ can be replaced by

$$u[-n-1] = \begin{cases} 1 & if \ -n-1 > 0 \\ 0 & otherwise \end{cases} \rightarrow u[-n-1] = \begin{cases} 1 & if \ n < -1 \\ 0 & otherwise \end{cases}$$

leading to the calculation

$$X(z) = \sum_{n=-\infty}^{-1} -\alpha^n z^{-n}$$

$$= -\sum_{n=1}^{\infty} \alpha^{-n} z^n$$

$$= 1 - \sum_{n=0}^{\infty} \alpha^{-n} z^n$$

$$= 1 - \frac{1}{1 - a^{-1}z} \quad |a^{-1}z| < 1 \rightarrow |z| < |a|$$

$$= \frac{1}{1 - a^{-1}z}.$$

Example 4.3 For $x[n] = \alpha^n u[n]$, find $X(z)$.

Solution 4.3 $X(z) = \frac{1}{1-a^{-1}z}$ ROC is $|z| > |a|$
The LTI system H given in Fig. 4.5.
Can be described as in Fig. 4.6.

Fig. 4.5 LTI system

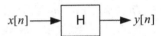

Fig. 4.6 LTI system with impulse response $h[n]$

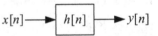

Fig. 4.7 LTI system with Z-transforms

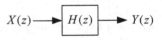

For the system of Fig. 4.6, $y[n] = h[n] * x[n]$ and we have $Y(z) = H(z)X(z)$. The LTI system H can also be described as in Fig. 4.7 using the Z-transforms.

Stability of a Discrete LTI System:

For a discrete LTI system to be a stable system, its impulse response should be absolutely summable, that is:

$$\sum_{n=-\infty}^{\infty} |h[n]| < \infty. \tag{4.14}$$

For a discrete LTI system, the transfer function is defined as

$$H(z) = \frac{Y(z)}{X(z)} \tag{4.15}$$

And for a discrete LTI system to be a stable system, poles of $H(z)$ should be inside the unit circle.

Example 4.4 For a discrete LTI system, the transfer function is given as

$$H(z) = \frac{z - 0.5}{(z - 0.3)(z - 0.8 - j0.8)}.$$

Determine whether the system is stable or not?

Solution 4.4 The poles of $H(z)$ are at $z_1 = 0.3$ and $z_2 = 0.8 + 0.8j$, and since $|z_2| = \sqrt{0.8^2 + 0.8^2} \rightarrow |z_2| = 1.13$ is outside the unit circle, the LTI system with the given transfer function is not a stable system.

4.1.2 Laplace Transform

Laplace transform is defined for continuous time signals. The Laplace transform of $h(t)$ is calculated as

$$H(s) = \int_{-\infty}^{\infty} h(t)e^{-st}dt \qquad (4.16)$$

where s is the complex frequency defined as $s = \sigma + jw$. The integral expression given in (4.16) converges for some set of s values which can be represented by a region in complex plane called convergence region or region of convergence in short.

The Properties of the Region of Convergence:

If $H(s) = \frac{P(s)}{Q(s)}$, then the roots of $P(s) = 0$ are called the zeros of $H(s)$ and the roots of $Q(s) = 0$ are called the poles of $H(s)$. The properties of the region of convergence (ROC) for $H(s)$ can be summarized as follows.

(1) The ROC does not include any poles.
(2) The ROC consists of vertical half planes or strips.
(3) Right side signals have ROC extending in the right half plane.
(4) Left side signals have ROC extending in the left half plane.
(5) Two sided signals do either have ROC in a central vertical strip or they diverge.

Stability of a Continuous LTI Systems:

The continuous LTI system H shown in Fig. 4.8.

Can also be described using its impulse response as in Fig. 4.9.

For the system of Fig. 4.9, we have $y_c(t) = h(t) * x_c(t)$ and $Y(s) = H(s)X(s)$. Thus, LTI system H can also be described as using Laplace transform of the functions as in Fig. 4.10.

In Fig. 4.10, $H(s)$ is called the transfer function of the continuous time system.

$$x_c(t) \longrightarrow \boxed{\text{H}} \longrightarrow y_c(t)$$

Fig. 4.8 A continuous LTI system

$$x_c(t) \longrightarrow \boxed{h_c(t)} \longrightarrow y_c(t)$$

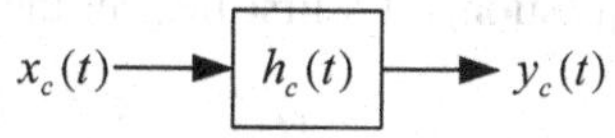

Fig. 4.9 A continuous LTI system with its impulse response

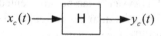

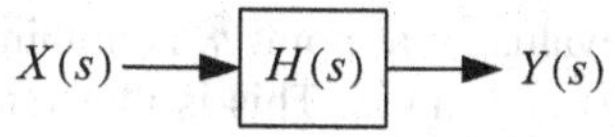

$$X(s) \longrightarrow \boxed{H(s)} \longrightarrow Y(s)$$

Fig. 4.10 A continuous LTI system using Laplace transforms

The continuous time system with impulse response $h(t)$ is stable if its impulse response is absolutely integrable, that is, continuous LTI system is stable if

$$\int_{\infty}^{\infty} |h(t)|dt < \infty. \tag{4.17}$$

If the transfer function $H(s)$ of the continuous time system is known, then the stability check can be performed by inspecting the poles of $H(s)$. If all the poles of $H(s)$ are in the left half plane, i.e., the complex poles have negative real parts, then the continuous time system is stable. Otherwise the system is unstable.

Example 4.5 For a continuous LTI system, the transfer function is given as

$$H(s) = \frac{s+1}{(s - 0.5 + 2j)(s + 3 - 2j)}.$$

Determine whether the system is stable or not?

Solution 4.5 The poles of $H(s)$ are $s_1 = 0.5 - 2j$ and $s_2 = -3 + 2j$. The system with transfer function $H(s)$ is not a stable system since the pole s_1 has positive real part.

For continuous LTI systems, the relationship between system input and system output can be described using differential equations as in

$$\sum_{k=0}^{N} a[k] \frac{d^k y(t)}{dt^k} = \sum_{k=0}^{M} b[k] \frac{d^k x(t)}{dt^k}. \tag{4.18}$$

4.2 Transformation Between Continuous and Discrete Time Systems

We know that continuous LTI systems can be represented by differential equations. And when the continuous time system is converted to a digital system, we can represent digital system by difference equations.

Now we ask the question: How can we convert a differential equation to a difference equation?

For the answer of this question, let's first inspect the conversion of

$$\frac{dx_c(t)}{dt}$$

to its discrete equivalent.

The derivative of $x_c(t)$ evaluated at point t_0 is nothing but the slope of the line tangent to the graph of $x_c(t)$ at point t_0. This is illustrated in the Fig. 4.11.

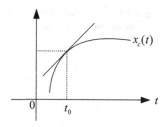

Fig. 4.11 A tangent line at point t_0

Now let's consider the digital signal obtained from $x_c(t)$ after sampling operation. The slope of the line tangent to the graph of $x_c(t)$ at point t_0 can be approximated using the sample values and sampling instants. The sampling of the continuous time signal is illustrated in the Fig. 4.12.

The slope of the line at point $t_0 = nT_s$ can be calculated using the triangles as shown in the Fig. 4.13.

The slope of the line tangent to the graph at point $t_0 = nT_s$ can be evaluated using the left triangle in Fig. 4.13 as

$$\frac{dx_c(t)}{dt}\bigg|_{t=nT_s} = \frac{x_c(nT_s) - x_c((n-1)T_s)}{T_s} \tag{4.19}$$

or using the right triangle in Fig. 4.13 as

$$\frac{dx_c(t)}{dt}\bigg|_{t=nT_s} = \frac{x_c((n+1)T_s) - x_c(nT_s)}{T_s}. \tag{4.20}$$

And we have the following identities

$$x[n] = x_c(nT_s) \quad x[n-1] = x_c((n-1)T_s) \quad x[n+1] = x_c((n+1)T_s). \tag{4.21}$$

Using (4.21) in (4.19) and (4.20), the derivative of the continuous time signal can be written either as

Fig. 4.12 Sampling of the continuous time signal

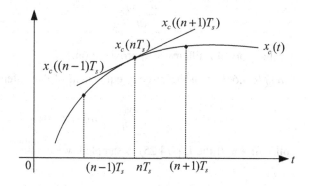

Fig. 4.13 Calculation of the slope of the tangent line at point $t_0 = nT_s$

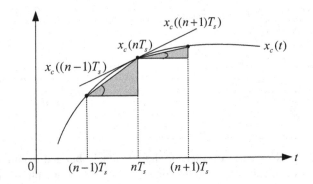

$$\left.\frac{dx_c(t)}{dt}\right|_{t=nT_s} \approx \frac{x[n] - x[n-1]}{T_s} \qquad (4.22)$$

or as

$$\left.\frac{dx_c(t)}{dt}\right|_{t=nT_s} \approx \frac{x[n+1] - x[n]}{T_s}. \qquad (4.23)$$

Otherwise indicated, we will use

$$\left.\frac{dx_c(t)}{dt}\right|_{t=nT_s} \approx \frac{x[n+1] - x[n]}{T_s} \qquad (4.24)$$

for the discrete approximation of the derivative operation.

In addition, the expression

$$\left.\frac{dx_c(t)}{dt}\right|_{t=nT_s} \approx \frac{x[n+1] - x[n]}{T_s}$$

is called backward difference approximation, and

$$\left.\frac{dx_c(t)}{dt}\right|_{t=nT_s} \approx \frac{x[n] - x[n-1]}{T_s}$$

is called forward difference approximation.

Example 4.6 Obtain the discrete equivalent of the differential equation

$$\frac{dy(t)}{dt} + ay(t) = bx(t). \qquad (4.25)$$

Solution 4.6 If the Eq. (4.25) is sampled, we obtain

$$\frac{dy(t)}{dt}\bigg|_{t=nT_s} + ay(t)|_{t=nT_s} = bx(t)|_{t=nT_s}. \tag{4.26}$$

And substituting

$$\frac{dy(t)}{dt}\bigg|_{t=nT_s} \approx \frac{y[n+1] - y[n]}{T_s}$$

$$y[n] = y(t)|_{t=nT_s}$$

$$x[n] = x(t)|_{t=nT_s} \tag{4.27}$$

into (4.26), we obtain the difference equation

$$\frac{y[n+1] - y[n]}{T_s} + ay[n] = bx[n]. \tag{4.28}$$

If we use the forward difference approximation

$$\frac{dy(t)}{dt}\bigg|_{t=nT_s} \approx \frac{y[n] - y[n-1]}{T_s}$$

we obtain

$$\frac{y[n] - y[n-1]}{T_s} + ay[n] = bx[n]$$

as the discrete approximation of (4.25).

Example 4.7 Find the discrete equivalent of

$$\frac{d^2y(t)}{dt^2}. \tag{4.29}$$

Solution 4.7 We can write

$$\frac{d^2y(t)}{dt^2}\bigg|_{t=nT_s}$$

as

$$\frac{d^2y(t)}{dt^2}\bigg|_{t=nT_s} = \frac{\frac{dy(t)}{dt}\big|_{t=(n+1)T_s} - \frac{dy(t)}{dt}\big|_{t=nT_s}}{T_s}. \tag{4.30}$$

Substituting

$$\frac{dy(t)}{dt}\bigg|_{t=nT_s} \approx \frac{y[n+1]-y[n]}{T_s}$$

into (4.30), we obtain

$$\frac{d^2y(t)}{dt^2}\bigg|_{t=nT_s} \approx \frac{y[n+2]-y[n+1]-(y[n+1]-y[n])}{T_s^2}$$

which can be simplified as

$$\frac{d^2y(t)}{dt^2}\bigg|_{t=nT_s} \approx \frac{y[n+2]-2y[n+1]+y[n]}{T_s^2}. \tag{4.31}$$

If we use forward difference approximation

$$\frac{dy(t)}{dt}\bigg|_{t=nT_s} \approx \frac{y[n]-y[n-1]}{T_s}$$

inside the expression

$$\frac{d^2y(t)}{dt^2}\bigg|_{t=nT_s} \approx \frac{\frac{dy(t)}{dt}\big|_{t=(nT_s)}-\frac{dy(t)}{dt}\big|_{t=(n-1)T_s}}{T_s}$$

we obtain

$$\frac{d^2y(t)}{dt^2}\bigg|_{t=nT_s} \approx \frac{\frac{y[n]-y[n-1]}{T_s}-\left(\frac{y[n-1]-y[n-2]}{T_s}\right)}{T_s}$$

which can be simplified as

$$\frac{d^2y(t)}{dt^2}\bigg|_{t=nT_s} \approx \frac{y[n]-2y[n-1]+y[n-2]}{T_s^2}. \tag{4.32}$$

Example 4.8 Find the discrete equivalent of the differential equation

$$\frac{d^2y(t)}{dt^2}+2\frac{dy(t)}{dt}+y(t)=x(t)+\frac{dx(t)}{dt}. \tag{4.33}$$

Solution 4.8 If both sides of the (4.33) are sampled, we get

$$\left.\frac{d^2y(t)}{dt^2}\right|_{t=nT_s} + 2\left.\frac{dy(t)}{dt}\right|_{t=nT_s} + y(t)|_{t=nT_s} = x(t)|_{t=nT_s} + \left.\frac{dx(t)}{dt}\right|_{t=nT_s}. \qquad (4.34)$$

And substituting the approximations and equations

$$\left.\frac{d^2y(t)}{dt^2}\right|_{t=nT_s} \approx \frac{y[n+2] - 2y[n+1] + y[n]}{T_s^2}$$

$$\left.\frac{dx(t)}{dt}\right|_{t=nT_s} \approx \frac{x[n+1] - x[n]}{T_s}$$

$$x[n] = x(t)|_{t=nT_s}$$

$$y[n] = y(t)|_{t=nT_s}$$

into (4.34), we obtain

$$\frac{y[n+2] - 2y[n+1] + y[n]}{T_s^2} + 2\frac{y[n+1] - y[n]}{T_s} + y[n] = x[n] + \frac{x[n+1] - x[n]}{T_s}. \qquad (4.35)$$

For $T_s = 1$, the Eq. (4.35) reduces to

$$y[n+2] = x[n+1].$$

Exercise: Find the discrete equivalent of

$$\frac{d^3y(t)}{dt^3}.$$

4.2.1 Conversion of Transfer Functions of LTI Systems

We know that continuous and discrete LTI systems can be described by differential or difference equations.

Fig. 4.14 Continuous time LTI system and its discrete equivalent

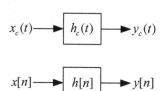

And a differential equation can be converted to a difference equation via sampling operation. The difference equation represents a discrete LTI system. In Fig. 4.14, a continuous time system and its discrete equivalent obtained via sampling operation is shown using block diagrams.

Both continuous and discrete systems have transfer functions defined as

$$H_c(s) = \frac{Y_c(s)}{X_c(s)} \quad \text{and} \quad H_n(z) = \frac{Y_n(z)}{X_n(z)}$$

respectively. Now we ask the question, given $H_c(s)$ can we obtain $H_n(z)$ from $H_c(s)$ directly?

The answer to this question is yes and we will derive two methods for the direct conversion of $H_c(s)$ to $H_n(z)$, and these methods will be called forward difference and bilinear transformation.

Note: For simplicity of notation, we will drop the subscript letters c and n from the equations $H_c(s)$ and $H_n(z)$.

4.2.2   Forward Difference Transformation Method

Consider the differential equation

$$\frac{dy(t)}{dt} + ay(t) = x(t) \tag{4.36}$$

which describes a continuous LTI system. Taking the Laplace transform of both sides of (4.36), we get

$$sY(s) + aY(s) = X(s)$$

from which the transfer function $H(s) = Y(s)/X(s)$ can be calculated as

$$H(s) = \frac{1}{s+a}. \tag{4.37}$$

If the differential equation

$$\frac{dy(t)}{dt} + ay(t) = x(t)$$

is sampled, we get

$$\frac{dy(t)}{dt}\bigg|_{t=nT_s} + ay(t)|_{t=nT_s} = x(t)|_{t=nT_s}$$

which yields the difference equation

$$\frac{y[n] - y[n-1]}{T_s} + ay[n] = x[n]. \tag{4.38}$$

And by taking the Z-transform of both sides of (4.38), we get

$$\frac{Y(z) - z^{-1}Y(z)}{T_s} + aY(z) = X(z)$$

from which the transfer function $H(z)$ can be calculated as

$$H(z) = \frac{1}{a + \frac{1-z^{-1}}{T_s}}. \tag{4.39}$$

When $H(s)$ in (4.37) and $H(z)$ in (4.39) are compared to each other as below

$$H(s) = \frac{1}{s+a} \quad H(z) = \frac{1}{a + \frac{1-z^{-1}}{T_s}}$$

we see that

$$H(z) = H(s)\big|_{s=\frac{1-z^{-1}}{T_s}} \tag{4.40}$$

Example 4.9 Obtain the discrete equivalent of

$$\frac{d^2y(t)}{dt^2} + \frac{dy(t)}{dt} + ay(t) = x(t) \tag{4.41}$$

and find the relation between $H(s)$ and $H(z)$. Use forward difference transformation method.

Solution 4.9 The discrete equivalent of

$$\frac{d^2y(t)}{dt^2} + \frac{dy(t)}{dt} + ay(t) = x(t) \tag{4.42}$$

is

$$\frac{y[n] - 2y[n-1] + y[n-2]}{T_s^2} + \frac{y[n] - y[n-1]}{T_s} + ay[n] = x[n]. \qquad (4.43)$$

Laplace transform of the (4.42) is

$$s^2 Y(s) + sY(s) + aY(s) = X(s). \qquad (4.44)$$

Z-transform difference Eq. (4.43) can be calculated as

$$\frac{Y(z) - 2z^{-1}Y(z) + z^{-2}Y(z)}{T_s^2} + \frac{Y(z) - z^{-1}Y(z)}{T_s} + aY(z) = X(z)$$

which yields

$$\left(\frac{1 - z^{-1}}{T_s}\right)^2 Y(z) - \frac{1 - z^{-1}}{T_s} Y(z) + aY(z) = X(z). \qquad (4.45)$$

If we compare the Laplace transform in (4.44) and Z-transform in (4.45), we see that Z-transform can be obtained from Laplace transform replacing s by $\frac{1-z^{-1}}{T_s}$. That is

$$H(z) = H(s)\big|_{s=\frac{1-z^{-1}}{T_s}} \qquad (4.46)$$

Therefore, if forward difference transformation method is used for any differential equation, the relation between transfer functions of continuous and discrete systems happens to be as in (4.46).

4.2.3 Bilinear Transformation

If the bilinear transformation method is used to obtain the difference equation from differential equation, the relation between transfer functions happens to be as

$$H(z) = H(s)\big|_{s=\frac{2}{T_s}\left(\frac{1-z^{-1}}{1+z^{-1}}\right)} \qquad (4.47)$$

Now let's derive the bilinear transformation formula in (4.47).
Consider the differential equation

$$\frac{dy(t)}{dt} + ay(t) = x(t). \qquad (4.48)$$

Let

$$w(t) = \frac{dy(t)}{dt}$$

then

$$y(t) = \int_{-\infty}^{t} w(\tau)d\tau$$

which can be written as

$$y(t) = \underbrace{\int_{-\infty}^{t_0} w(\tau)d\tau}_{y(t_0)} + \int_{t_0}^{t} w(\tau)d\tau$$

$$y(t) = y(t_0) + \int_{t_0}^{t} w(\tau)d\tau. \tag{4.49}$$

When the Eq. (4.49) is sampled at time instants $t = nT_s$ and $t_0 = (n-1)T_s$, we get

$$\underbrace{y(nT_s)}_{y[n]} = \underbrace{y((n-1)T_s)}_{y[n-1]} + \int_{(n-1)T_s}^{nT_s} w(\tau)d\tau \tag{4.50}$$

which can be written as

$$y[n] = y[n-1] + \int_{(n-1)T_s}^{nT_s} w(\tau)d\tau. \tag{4.51}$$

Now let's consider the evaluation of the integral expression in (4.51). We can evaluate the integration in (4.51) using the trapezoidal integration rule. This is shown in the Fig. 4.15.

Using Fig. 4.15, we can write

$$\int_{(n-1)T_s}^{nT_s} w(\tau)d\tau = \frac{T_s}{2}\left(w((n-1)T_s) + w(nT_s)\right) \tag{4.52}$$

Fig. 4.15 Trapezoidal
integration

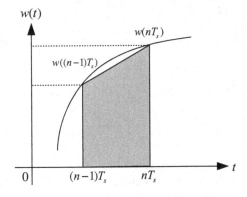

which can be simplified as

$$\int_{(n-1)T_s}^{nT_s} w(\tau)d\tau = \frac{T_s}{2}(w[n-1]+w[n]).$$ (4.53)

Substituting (4.53) into (4.51), we obtain

$$y[n] = y[n-1] + \frac{T_s}{2}(w[n-1]+w[n]).$$ (4.54)

Consider the equation

$$\underbrace{\frac{dy(t)}{dt}}_{w(t)} + ay(t) = x(t).$$ (4.55)

When (4.55) is sampled, we obtain

$$w[n] + ay[n] = x[n] \rightarrow w[n] = -ay[n] + x[n].$$ (4.56)

If Eq. (4.56) is substituted into (4.54), we obtain

$$y[n] = y[n-1] + \frac{T_s}{2}(-ay[n-1]+x[n-1]-ay[n]+x[n])$$ (4.57)

which can be rearranged as

$$y[n] + \frac{aT_s}{2}y[n] + \frac{aT_s}{2}y[n-1] - y[n-1] = +\frac{T_s}{2}x[n-1] + \frac{T_s}{2}x[n].$$ (4.58)

And taking the Z-transform of both sides of (4.58), we get

$$\left(1 + \frac{aT_s}{2}\right)Y(z) - \left(1 - \frac{aT_s}{2}\right)z^{-1}Y(z) = \frac{T_s}{2}\left(a + z^{-1}\right)X(z) \tag{4.59}$$

from which the transfer function can be calculated as

$$H(z) = \frac{Y(z)}{X(z)} \rightarrow H(z) = \frac{1}{a + \frac{2}{T_s}\left(\frac{1-z^{-1}}{1+z^{-1}}\right)}. \tag{4.60}$$

When (4.60) is compared to

$$H(s) = \frac{1}{a + s} \tag{4.61}$$

we see that

$$H(z) = H(s)\big|_{s = \frac{2}{T_s}\left(\frac{1-z^{-1}}{1+z^{-1}}\right)} \tag{4.62}$$

Bilinear transformation is an efficient transformation technique. Stable continuous time LTI systems are converted into stable discrete LTI systems.

That is if the poles of $H(s)$ are in the left half plane, the poles of $H(z)$ are inside the unit circle. This is illustrated in Fig. 4.16.

Frequency Mapping in Bilinear Transformation:

In bilinear transformation, the relation between continuous and digital frequency is given as

$$s = \frac{2}{T_s}\left(\frac{1 - z^{-1}}{1 + z^{-1}}\right) \tag{4.63}$$

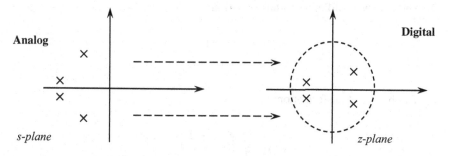

Fig. 4.16 Pole mapping in bilinear transformation

where $s = \sigma + jw_a$ and $z = e^{jw_d}$. Let

$$w_a \rightarrow \text{Analog signal frequency}$$

and

$$w_d \rightarrow \text{Digital signal frequency.}$$

Equation (4.63) yields

$$
\begin{aligned}
\sigma + jw_a &= \frac{2}{T_s}\left(\frac{1 - e^{-jw_d}}{1 + e^{-jw_d}}\right) \\
&= \frac{2}{T_s}\left(\frac{e^{-j\frac{w_d}{2}}\left(e^{j\frac{w_d}{2}} - e^{-j\frac{w_d}{2}}\right)}{e^{-j\frac{w_d}{2}}\left(e^{j\frac{w_d}{2}} + e^{-j\frac{w_d}{2}}\right)}\right) \\
&= j\frac{2}{T_s}\frac{\sin\left(\frac{w_d}{2}\right)}{\cos\left(\frac{w_d}{2}\right)} \\
&= j\frac{2}{T_s}\tan\left(\frac{w_d}{2}\right).
\end{aligned}
$$

Hence,

$$w_a = \frac{2}{T_s}\tan\left(\frac{w_d}{2}\right). \tag{4.64}$$

Summary: Transformation of analog systems to discrete ones can be achieved by using the following methods.

(1) The forwards difference transformation:

$$s = \frac{z - 1}{T_s}.$$

(2) The backward difference transformation:

$$s = \frac{z - 1}{T_s z}.$$

(3) The bilinear transformation:

$$s = \frac{2}{T_s}\left(\frac{1 - z^{-1}}{1 + z^{-1}}\right)$$

(4) Impulse invariance transformation:

$$H(z) = T_s \times Z-\text{transform of } \{H(s)\}.$$

(5) Step invariance transformation:

$$H(z) = \left(1 - z^{-1}\right) \times Z-\text{transform of } \left\{\frac{H(s)}{s}\right\}.$$

Example 4.10 Transfer function of a continuous time system is given as

$$H(s) = \frac{4s + 11}{s^2 + 7s + 10}.$$

Find the transfer function $H(z)$ of the digital system obtained via the sampling of continuous time system.

Solution 4.10 $H(z) = H(s)|_{s=\frac{2}{T_s}\left(\frac{1-z^{-1}}{1+z^{-1}}\right)}$, for simplicity of the calculation, we can choose $T_s = 1$ and this yields

$$H(z) = \frac{19 + 22z^{-1} + 3z^{-2}}{28 + 12z^{-1}}.$$

4.3 Analogue Filter Design

Consider the continuous LTI system given in Fig. 4.17.
Where the system output equals to

$$y(t) = x(t) * h(t)$$

which can be written in frequency domain as

$$Y(w) = X(w)H(w). \tag{4.65}$$

If the magnitude of $H(w)$ in (4.65) gets very small values for some specific values of w, the output function $Y(w)$ does no contain any information about $X(w)$ and this operation is called filtering.

Fig. 4.17 A continuous LTI system

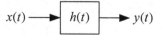

Any analog filter is characterized by its transfer function $H(w)$ which can be a complex function with magnitude $|H(w)|$ and phase $\angle H(w)$ characteristics.

If we denote the phase characteristics as

$$\theta(w) = \angle H(w) \rightarrow \theta(w) = \arg(H(w))$$

then phase and group delays are defined as

$$\rho(w) = -\frac{\theta(w)}{dw} \quad \tau(w) = -\frac{d\theta(w)}{dw}. \tag{4.66}$$

Group delay function gives information about the amount of delay introduced by the system transfer function to the system input. For instance, if

$$\tau(w) = 2$$

then for the transfer function with unit gain the system input

$$x(t) = \sin(wt)$$

yields the system output

$$y(t) = \sin(w(t - 2)).$$

4.3.1   Ideal Filters

In this section we will study the transfer functions of the ideal filters. For $H(w)$, i.e., the transfer function of the ideal filter, the time domain impulse response can be calculated using the inverse Fourier transform

$$h(t) = \frac{1}{2\pi} \int_{w=-\infty}^{\infty} H(w)dw$$

which is a function having non-zero values for all t values in the range $-\infty < t < \infty$, for this reason such filters are not physically realizable, and they are called ideal filters.

Ideal Low-Pass Filter:

The transfer function of the ideal low-pass filter is shown in Fig. 4.18.

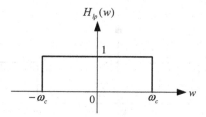

Fig. 4.18 Transfer function of the ideal low-pass filter

Whose impulse response can be calculated as

$$h_{lp}(t) = \frac{1}{2\pi} \int_{-w_c}^{w_c} 1 \times e^{jwt} dw$$

$$= \frac{1}{\pi t} \sin c(w_c t)$$

where w_c is called cut-off frequency.

Ideal High-Pass Filter:

The transfer function of the ideal high-pass filter is shown in Fig. 4.19.

Which can be written in terms of the transfer function of the low-pass filter with the same cut-off frequency as

$$H_{hp}(w) = 1 - H_{lp}(w). \tag{4.67}$$

whose inverse Fourier transform equals to

$$h_{hp}(t) = 1 - \frac{1}{\pi t} \sin c(w_c t). \tag{4.68}$$

Ideal Band-Pass Filter:

The transfer function of the ideal band-pass filter is shown in Fig. 4.20.

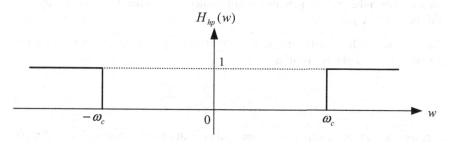

Fig. 4.19 Transfer function of the ideal high-pass filter

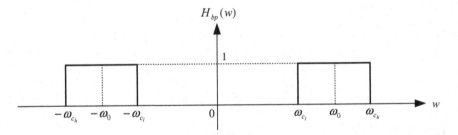

Fig. 4.20 Transfer function of the ideal band-pass filter

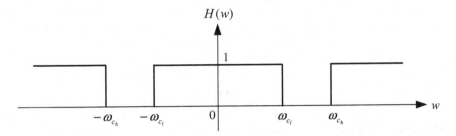

Fig. 4.21 Transfer function of the ideal band-stop filter

Which can be obtained from low-pass filter transfer function with the same cut-off frequency as

$$H_{bp}(w) = H_{lp}(w - w_0) + H_{lp}(w + w_0). \tag{4.69}$$

In Fig. 4.20; w_{c_l} and w_{c_h} are low and high cut-off frequencies.
Ideal Band-Stop Filter:
The transfer function of the ideal band-stop filter is shown in Fig. 4.21.
Which can be obtained from band-pass filter transfer function (4.69) as

$$H_{bs}(w) = 1 - H_{bp}(w). \tag{4.70}$$

As can be seen from the filter transfer functions; if we design a low-pass filter, we can obtain the transfer function of other filters by just manipulating the transfer function of low-pass filter.

Example 4.11 The transfer function of an analog low-pass filter with cut-off frequency $\omega_c = 1$ rad/s is given as

$$H_1(w) = \frac{1}{w^2 + 2\sqrt{2}w + 4}.$$

Find the transfer function of low-pass filter with cut-off frequency $\omega_c = 2$ rad/s.

Fig. 4.22 Transfer function of the ideal low-pass filter with cut-off frequency $\omega_c = 1$ rad/s

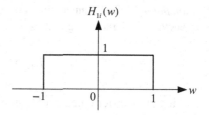

Fig. 4.23 Transfer function of the ideal low-pass filter with cut-off frequency $\omega_c = 2$ rad/s

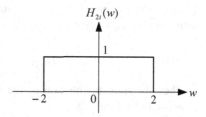

Solution 4.11 The transfer function of the ideal low-pass filter with cut-off frequency $\omega_c = 1$ rad/s is shown in the Fig. 4.21.

And the transfer function of the ideal low-pass filter with cut-off frequency $\omega_c = 2$ rad/s is shown in the Fig. 4.4.

From Figs. 4.22 and 4.23, we see that

$$H_{2i}(w) = H_{1i}\left(\frac{w}{2}\right) \tag{4.71}$$

In a similar manner, using the low-pass filter with cut-off frequency $\omega_c = 1$ rad/s in the problem, we can calculate the transfer function of the low-pass filter with cut-off frequency $\omega_c = 2$ rad/s employing (4.71) as

$$H_2(w) = \frac{4}{w^2 + 4\sqrt{2}w + 16}.$$

In general, given the transfer function of low-pass filter $H_1(w)$ with cut-off frequency 1 rad/s, the transfer function of low-pass filter with cut-off frequency ω_c can be obtained as

$$H_{w_c}(w) = H_1\left(\frac{w}{w_c}\right) \tag{4.72}$$

Example 4.12 The transfer function of an analog low-pass filter with cut-off frequency $\omega_c = 1$ rad/s is given as

$$H_1(s) = \frac{1}{s^2 + 2\sqrt{2} + 4}.$$

Find the transfer function of high-pass filter with cut-off frequency $\omega_c = 2$ rad/s.

Solution 4.12 First, we can design the low-pass filter with cut-off frequency $\omega_c = 2$ rad/s as in the previous example and the transfer function of the low-pass filter with cut-off frequency $\omega_c = 2$ rad/s is found as

$$H_{lp}(w) = \frac{4}{w^2 + 4\sqrt{2}w + 16}.$$

Then the transfer function of the high-pass filter with cut-off frequency $\omega_c = 2$ rad/s can be found as

$$\begin{aligned} H_{hp} &= 1 - H_{lp}(w) \\ &= \frac{w^2 + 4\sqrt{2}w + 12}{w^2 + 4\sqrt{2}w + 16}. \end{aligned}$$

Hence, for the filter design; it is custom to design a low-pass filter with cut-off frequency $\omega_c = 1$ rad/s and transfer it to any desired frequency response.

4.3.2   Practical Analog Filter Design

Although ideal filters are simple to understand they cannot be used to construct filter circuits; since they need an infinite number of circuit elements. For this reason, practical analog filter design techniques are adapted in the signal processing literature. The specifications of a practical analog filter are given in Fig. 4.24.

Fig. 4.24 The specifications of a practical analog filter

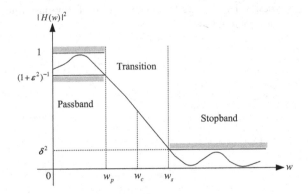

As can be seen from Fig. 4.24, the squared filter magnitude should satisfy

$$\left(1+\epsilon^2\right)^{-1} \leq |H(w)|^2 \leq 1 \quad \text{for} \quad 0 \leq w \leq w_p$$

in passband and it should satisfy

$$0 \leq |H(w)|^2 \leq \delta^2 \quad \text{for} \quad w_s \leq w \leq \infty$$

in stopband.

Filter Parameters

Cut-off frequency:

At cut-off frequency w_c, the amplitude of the transfer function equals to $\frac{1}{\sqrt{2}}|H(w)|_{max}$, that is

$$H(w_c) = \frac{1}{\sqrt{2}}|H(w)|_{max}.$$

If $|H(w)|_{max} = 1$, then w_c is determined from

$$H(w_c) = \frac{1}{\sqrt{2}}.$$

Pass-band ripple:

Passband ripple in decibels is defined as

$$R_p = 10\log\left(1+\epsilon^2\right). \tag{4.73}$$

Stopband attenuation:

The stopband attenuation is defined as

$$R_s = -10\log\left(\delta^2\right). \tag{4.74}$$

Selectivity parameter:

The ratio of pass-band frequency to stop-band frequency is called selectivity parameters, i.e.,

$$k = \frac{w_p}{w_s}$$

which is equal to 1 for ideal filters, and for practical filters $k < 1$.

Discrimination parameter:
The discrimination parameter is used as an indicator of the pass-band and stop-band attenuation ratios and defined as

$$d = \frac{\epsilon}{\sqrt{\delta^{-2} - 1}}$$

which is equal to 0 for ideal filters and $d > 1$ for practical filters.
Now let's see the practical filter design methods.

4.3.3 Practical Filter Design Methods

The most known practical filter design techniques in literature are:

(1) Butterworth filter design.
(2) Chebyshev I and II filter design.
(3) Elliptic filter design.
(4) Bessel filter design.

4.3.3.1 Butterworth Filter Design

The squared magnitude response of the Nth order Butterworth filter is defined as

$$|H(w)|^2 = \frac{1}{1 + \left(\frac{w}{w_c}\right)^{2N}} \qquad (4.75)$$

where w_c is the cut-off frequency.
The transfer function of the Nth order Butterworth filter is

$$H(s) = \frac{w_c^N}{\prod_{k=1}^{N}(s - p_k)} \qquad (4.76)$$

where the poles p_k are given as

$$p_k = w_c e^{\frac{j\pi}{2}\left(1 + \left(\frac{2k-1}{N}\right)\right)}. \qquad (4.77)$$

The transfer function $H(s)$ has N poles located on a circle of radius w_c on the left half plane.

Given low-pass filter specifications w_p, w_s, R_p, the low-pass Butterworth filter is designed via the following steps:

(1) Using the given filter specifications and the expression

$$|H(w)|^2 = \frac{1}{1 + \left(\frac{w}{w_c}\right)^{2N}}$$

decide on the filter order N and cut-off frequency w_c.

(2) Determine the poles using

$$p_k = w_c e^{\frac{j\pi}{2}\left(1 + \left(\frac{2k-1}{N}\right)\right)}, \quad k = 1, \ldots, N.$$

(3) Find the transfer function using the poles as

$$H(s) = \frac{w_c^N}{\prod_{k=1}^{N}(s - p_k)}.$$

(4) And finally construct the filter circuit using the transfer function $H(s)$ found in the previous step.

Filter order N and cut-off frequency determination:

(a) From Fig. 4.24, we see that

$$\text{at } w = w_p \quad |H(w_p)| = \frac{1}{1 + \left(\frac{w_p}{w_c}\right)^{2N}} \rightarrow |H(w_p)| = \frac{1}{1 + \epsilon^2}$$

which leads to the equation

$$\frac{1}{1 + \left(\frac{w_p}{w_c}\right)^{2N}} = \frac{1}{1 + \epsilon^2}. \tag{4.78}$$

(b) In a similar manner, from Fig. 4.24, it is also seen that

$$\text{at } w = w_s \quad |H(w_s)| = \frac{1}{1 + \left(\frac{w_s}{w_c}\right)^{2N}} \rightarrow |H(w_p)| = \delta^2$$

which yields the equation

$$\frac{1}{1+\left(\frac{w_p}{w_c}\right)^{2N}} = \delta^2. \tag{4.79}$$

From (4.78) and (4.79), we obtain the equation set

$$\left.\begin{array}{l}\left(\dfrac{w_s}{w_c}\right)^{2N} = \delta^{-2} - 1 \\[2mm] \left(\dfrac{w_p}{w_c}\right)^{2N} = 1 + \epsilon^2\end{array}\right\} \rightarrow \text{dividing them}$$

we get

$$\left(\frac{w_s}{w_p}\right)^{2N} = \frac{\delta^{-2} - 1}{\epsilon^2}. \tag{4.80}$$

When (4.80) is solved for N, we get

$$N \geq \left\lceil \frac{\log\left(\frac{\sqrt{\delta^{-2}-1}}{\epsilon}\right)}{\log\left(\frac{w_s}{w_p}\right)} \right\rceil \tag{4.81}$$

which can be written in terms of selectivity and discrimination parameters as

$$N \geq \left\lceil \frac{\log\left(\frac{1}{d}\right)}{\log\left(\frac{1}{k}\right)} \right\rceil \tag{4.82}$$

where $\lceil \cdot \rceil$ is the round up to the larger integer function.

And the cut-off frequency w_c can be determined by solving one of the equations

$$\begin{array}{l}\left(\dfrac{w_s}{w_c}\right)^{2N} = \delta^{-2} - 1 \\[3mm] \left(\dfrac{w_p}{w_c}\right)^{2N} = 1 + \epsilon^2\end{array} \tag{4.83}$$

yielding the roots

$$w_c = \epsilon^{-\frac{1}{N}} w_p \qquad w_c = \left(\delta^{-2} - 1\right)^{-\frac{1}{2N}} w_s. \tag{4.84}$$

Or the cut-off frequency can be selected as any value from the range

$$\epsilon^{-\frac{1}{N}} w_p \leq w_c \leq \left(\delta^{-2} - 1\right)^{-\frac{1}{2N}} w_s. \tag{4.85}$$

Example 4.13 Design the transfer function of low-pass Butterworth filter whose specifications are given as

$$w_p = 1000 \text{ rad/s} \quad w_s = 3000 \text{ rad/s} \quad R_p = 4 \text{ dB} \quad R_s = 40 \text{ dB}.$$

Solution 4.13 Let's first determine the ϵ and δ values using R_p and R_s given in the question as follows

$$R_p = 10 \log\left(1 + \epsilon^2\right) \rightarrow 4 = 10 \log\left(1 + \epsilon^2\right) \rightarrow \epsilon^2 = 1.51 \rightarrow \epsilon = 1.23$$
$$R_s = -10 \log\left(\delta^2\right) \rightarrow 40 = -10 \log\left(\delta^2\right) \rightarrow \delta^2 = 10^{-4}.$$

And using the calculated ϵ^2 and δ^2 values in the Fig. 4.25.
We can roughly sketch the filter squared magnitude response as in Fig. 4.26.
Next, we determine the order N of the filter as follows

$$k = \frac{w_p}{w_s} \rightarrow k = \frac{1}{3}$$
$$d = \frac{\epsilon}{\sqrt{\delta^{-2} - 1}} \rightarrow d = \frac{1.23}{\sqrt{10^4 - 1}} \rightarrow d \approx 0.0123$$
$$N \geq \frac{\log\left(\frac{1}{d}\right)}{\log\left(\frac{1}{k}\right)} \rightarrow N \geq \frac{\log\left(\frac{1}{0.0123}\right)}{\log(3)} = 4.002 \rightarrow N = 4.$$

And the cut-off frequency can be found using

$$\epsilon^{-\frac{1}{N}} w_p \leq w_c \leq \left(\delta^{-2} - 1\right)^{-\frac{1}{2N}} w_s$$

Fig. 4.25 Typical magnitude squared transfer function of a practical low-pass filter

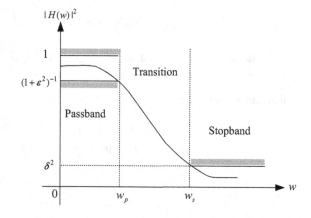

Fig. 4.26 Magnitude squared transfer function of a practical low-pass filter for Example 4.13

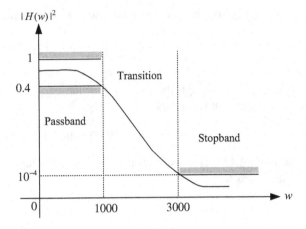

as follows

$$1.23^{-\frac{1}{4}}1000 \leq w_c \leq \left(10^4 - 1\right)^{-\frac{1}{8}}3000$$

$$949.6 \leq w_c \leq 948.69 \rightarrow w_c = 949 \text{ rad/s}.$$

The poles for $N = 4$ are calculated using

$$p_k = w_c e^{\frac{j\pi}{2}\left(1 + \left(\frac{2k-1}{N}\right)\right)}, \quad k = 1, \ldots, N$$

as follows

$$p_1 = 949e^{\frac{j\pi}{2}\left(1+\frac{1}{4}\right)} \rightarrow p_1 = 949e^{\frac{j5\pi}{8}} \rightarrow p_1 = 949\left(\cos\left(\frac{5\pi}{8}\right) + j\sin\left(\frac{5\pi}{8}\right)\right)$$

$$p_2 = 949e^{\frac{j\pi}{2}\left(1+\frac{3}{4}\right)} \rightarrow p_2 = 949e^{\frac{j7\pi}{8}} \rightarrow p_2 = 949\left(\cos\left(\frac{7\pi}{8}\right) + j\sin\left(\frac{7\pi}{8}\right)\right)$$

$$p_3 = 949e^{\frac{j\pi}{2}\left(1+\frac{5}{4}\right)} \rightarrow p_3 = 949e^{\frac{j9\pi}{8}} \rightarrow p_3 = 949\left(\cos\left(\frac{9\pi}{8}\right) + j\sin\left(\frac{9\pi}{8}\right)\right)$$

$$p_4 = 949e^{\frac{j\pi}{2}\left(1+\frac{7}{4}\right)} \rightarrow p_4 = 949e^{\frac{j11\pi}{8}} \rightarrow p_4 = 949\left(\cos\left(\frac{11\pi}{8}\right) + j\sin\left(\frac{11\pi}{8}\right)\right).$$

which can be simplified as

$$p_1 = -363 + 876j \quad p_2 = -876 + 363j$$
$$p_3 = -876 - 363j \quad p_4 = -363 - 876j.$$

Using the calculated poles, the transfer function is evaluated as

$$H(s) = \frac{w_c^N}{(s - p_1)(s - p_2)(s - p_3)(s - p_4)}$$

which leads to the expression

$$H(s) = \frac{949^2}{\left((s + 363)^2 - 876^2\right)\left((s + 876)^2 - 363^2\right)}$$

whose simplified form is

$$H(s) = \frac{900{,}601}{(s^2 + 726s - 635{,}607)(s^2 + 1752s + 635{,}607)}.$$

4.3.3.2 Chebyshev Filter Design

Chebyshev Type-I Filter:
Chebyshev Type-I filter squared magnitude response is equiripple in the passband and monotonic in the stopband. The squared magnitude response of a typical Chebyshev Type-I filter is depicted in the Fig. 4.27.

In Chebyshev Type-I filter transition from passband to stopband is more rapid when compared to Butterworth filter.

The square magnitude response of Chebyshev Type-I filter is defined as

$$|H_I(w)|^2 = \frac{1}{1 + \epsilon^2 T_N^2\left(\frac{w}{w_p}\right)} \tag{4.86}$$

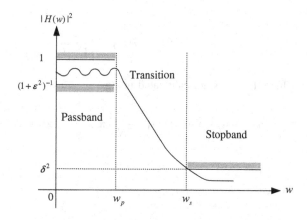

Fig. 4.27 Square magnitude response of Chebyshev Type-I filter

where $T_N(w)$ is the Nth order Chebyshev polynomial given as

$$T_N(w) = \begin{cases} \cos(N\cos^{-1}(w)) & |w| \le 1 \\ \cos h(N\cos h^{-1}(w)) & |w| > 1 \end{cases}. \tag{4.87}$$

The Chebyshev polynomial can be calculated in an iterative manner as

$$T_m(w) = 2wT_{m-1}(w) - T_{m-2}(w) \quad m \ge 2 \tag{4.88}$$

with the initial conditions

$$T_0(w) = 1 \quad \text{and} \quad T_1(w) = w. \tag{4.89}$$

Chebyshev Type-I filter design:
Assume that the low-pass filter specifications w_p, w_s, R_p, R_s are given. The design of the Chebychev Type-I filter can be achieved via the following steps

(1) First, with the given low-pass filter specifications; the order of the filter is determined as:

$$N \ge \frac{\log\left(d^{-1} + \sqrt{d^{-2} - 1}\right)}{\log\left(k^{-1} + \sqrt{k^{-2} - 1}\right)} = \frac{\cos h^{-1}(d^{-1})}{\cos h^{-1}(k^{-1})} \tag{4.90}$$

where k and d are the selectivity and discrimination parameters, and R_p is the passband ripple. The cut-off frequency is found by solving the equation

$$H(w_c) = 10^{-\frac{R_p}{10}}. \tag{4.91}$$

(2) Next, we calculate the transfer function

$$H(s) = \frac{c}{\prod_{k=1}^{N}(s - p_k)} \tag{4.92}$$

where the poles are calculated using

$$p_k = -w_p \sin h(\phi) \sin\left(\frac{2k-1}{2N}\pi\right) + jw_p \cos h(\phi) \cos\left(\frac{2k-1}{2N}\pi\right) \tag{4.93}$$

in which ϕ is defined as

$$\phi = \frac{1}{N}\ln\left(\frac{1+(1+\epsilon^2)^{\frac{1}{2}}}{\epsilon}\right). \tag{4.94}$$

And the constant term c in (4.92) is calculated via

$$c = \begin{cases} -\displaystyle\prod_{k=1}^{N} p_k & \text{if } N \text{ is odd} \\[2ex] (1+\epsilon^2)^{-\frac{1}{2}}\displaystyle\prod_{k=1}^{N} p_k & \text{if } N \text{ is even} \end{cases} \tag{4.95}$$

Example 4.14 Design a low-pass filter whose specifications are given as

$$w_p = 1000 \text{ rad/s} \quad w_s = 4000 \text{ rad/s} \quad R_p = 5 \text{ dB} \quad R_s = 40 \text{ dB}.$$

Use the transfer function of Chebyshev Type-I filter for your design.

Solution 4.14 With the given filter specifications, the parameters ϵ and δ are calculated as

$$R_p = 10\log(1+\epsilon^2) \rightarrow 5 = 10\log(1+\epsilon^2) \rightarrow \epsilon^2 = 2.16 \rightarrow \epsilon = 1.47$$
$$R_s = -10\log(\delta^2) \rightarrow 40 = -10\log(\delta^2) \rightarrow \delta^2 = 10^{-4} \rightarrow \delta = 10^{-2}.$$

And selectivity and discrimination parameters are found via

$$k = \frac{w_p}{w_s} \rightarrow k = \tfrac{1}{4}$$
$$d = \frac{\epsilon}{\sqrt{\delta^{-2}-1}} \rightarrow d = \frac{1.47}{\sqrt{10^4-1}} \rightarrow d \approx 0.0147.$$

The filter order is calculated as

$$N \geq \frac{\cos\text{h}^{-1}(d^{-1})}{\cos\text{h}^{-1}(k^{-1})} \rightarrow N \geq 2.38 \rightarrow N = 3.$$

The calculation of the poles can be achieved via

$$\phi = \frac{1}{N}\ln\left(\frac{1+(1+\epsilon^2)^{\frac{1}{2}}}{\epsilon}\right) \rightarrow \phi = 0.2121$$

$$p_k = -w_p \sin h(\phi) \sin\left(\frac{2k-1}{2N}\pi\right) + jw_p\cos h(\phi)\cos\left(\frac{2k-1}{2N}\pi\right)$$

$$p_1 = -1000\sin h(0.2121)\sin\left(\frac{\pi}{6}\right) + j1000\cos h(0.2121)\cos\left(\frac{\pi}{6}\right)$$

$$p_1 = -106.8 + 885.5j$$

$$p_2 = -1000\sin h(0.2121)\sin\left(\frac{3\pi}{6}\right) + j1000\cos h(0.2121)\cos\left(\frac{3\pi}{6}\right)$$

$$p_2 = -213.7$$

$$p_3 = -1000\sin h(0.2121)\sin\left(\frac{5\pi}{6}\right) + j1000\cos h(0.2121)\cos\left(\frac{5\pi}{6}\right)$$

$$p_3 = -106.8 - 885.5j.$$

Since N is odd, the constant term is calculated using

$$c = -\prod_{k=1}^{3} p_k \rightarrow c = 170{,}040{,}000.$$

Then the transfer function of the filter is calculated via

$$H(s) = \frac{c}{\prod_{k=1}^{N}(s - p_k)}$$

leading to the expression

$$H(s) = \frac{170{,}040{,}000}{(s+213.7)((s+106.8)^2 + 885.5^2)}$$

which can be simplified as

$$H(s) = \frac{170{,}040{,}000}{(s+213.7)(s^2 + 213.6s + 784{,}110)}.$$

And the above transfer function can be implemented using operational amplifiers and passive circuit elements.

Chebyshev Type-II Filter:

Chebyshev Type-II filter's magnitude squared response is monotonic in the passband and equiripple is the stopband. The magnitude squared response of a typical Chebyshev Type-II filter is depicted in the Fig. 4.28.

The magnitude squared response of Type-II Chebyshev filter can be given in two different forms as

Fig. 4.28 The magnitude squared response of a typical Chebyshev Type-II filter

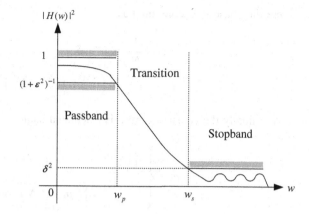

$$|H_{II}(w)|^2 = \frac{\epsilon^2 T_N^2\left(\frac{w_s}{w}\right)}{1 + \epsilon^2 T_N^2\left(\frac{w_s}{w}\right)}$$

$$|H_I(w)|^2 = \frac{1}{1 + \epsilon^2 T_N^2\left(\frac{w}{w_p}\right)}. \tag{4.96}$$

The relationship between the two transfer functions in (4.96) is given as

$$|H_{II}(w)|^2 = 1 - \left|H_I\left(\frac{1}{w}\right)\right|^2 \qquad w_p = \frac{1}{w_s}. \tag{4.97}$$

The transfer function of the Type-II Chebyshev filter is defined as

$$H(s) = \begin{cases} c \prod_{k=1}^{N} \frac{s-z_i}{s-p_i} & \text{if } N \text{ is even} \\[2mm] \left(\frac{c}{s - p_{\frac{N+1}{2}}}\right) \prod_{\substack{k=1 \\ k \neq \frac{N+1}{2}}}^{N} \frac{s-z_i}{s-p_i} & \text{if } N \text{ is even} \end{cases} \tag{4.98}$$

where z_i and p_i are the zeros and poles of the transfer function and they are calculated using

$$z_i = j\frac{w_s}{\cos\left(\frac{2k-1}{2N}\right)\pi} \tag{4.99}$$

$$p_i = \frac{w_s}{\alpha_i^2 + \beta_i^2}\left(-\sin h(\phi)\sin\left(\frac{2k-1}{2N}\pi\right) + j\cos h(\phi)\cos\left(\frac{2k-1}{2N}\pi\right)\right) \tag{4.100}$$

where the phase ϕ is computed as

$$\begin{aligned}
\phi &= \frac{1}{N}\cos h^{-1}\left(\delta^{-1}\right) \\
&= \frac{1}{N}\ln\left(\delta^{-1} + \left(\delta^{-2} - 1\right)^{\frac{1}{2}}\right).
\end{aligned} \tag{4.101}$$

And finally the constant term c is calculated using

$$c = \begin{cases}
\displaystyle\prod_{k=1}^{N}\frac{p_k}{z_k} & \text{if } N \text{ is even} \\[2ex]
\displaystyle\left(1+\epsilon^2\right)^{-\frac{1}{2}}\prod_{k=1}^{N}p_k & \text{if } N \text{ is odd.}
\end{cases}$$

4.3.3.3   Elliptic Filters

The magnitude squared response of the elliptic filters are given as

$$|H(w)|^2 = \frac{1}{1 + \epsilon^2 U_N^2(w)}$$

where $U_N(w)$ is the Jacobian elliptic function.

Elliptic filters have equiripple both in the passband and stopband. The amount of the ripple in each band can be adjusted. When the ripple in stopband approaches to zero, the filter converged to a Type-I Chebyshev filter. On the other hand, as the ripple in passband approaches to zero, the filter converged to a Type-II Chebyshev filter. If the ripples in both bands approaches to zero, then the filter converged to a Butterworth filter.

Elliptic filters have the steepest roll-off characteristics. The squared magnitude response of a typical Elliptic filter is depicted in the Fig. 4.29.

Fig. 4.29 The squared magnitude response of a typical Elliptic filter

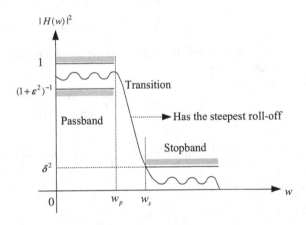

The phase response of the Elliptic filters is a non-linear function. The design of the elliptic filters is relatively complex when compared to Butterworth and Chebyshev filters.

4.3.3.4 Bessel Filters

For Butterworth, Chebyshev and Elliptic filters; the group delay $\tau(\theta)$ is a nonlinear function of the frequency. This means that the time delay introduced to the system varies nonlinearly with the frequency.

Bessel filters are linear phase filters and the group delay for these filters is a constant number independent of the frequency. For this reason, a constant time delay is introduced into the system independent of the frequency.

However, Bessel filters has the lowest roll-off factor among all the practical filters we have mentioned up to now. The squared magnitude response of a typical Bessel filter is depicted in the Fig. 4.30.

Summary:

Butterworth Filters: No ripple in passband and stopband. Group delay is nonlinear function of the frequency. Roll-off is low.

Chebyshev Type-I Filters: Have ripple in passband, no ripple in stopband. Group delay is a nonlinear function of the frequency. Roll-off is high.

Chebyshev Type-II Filters: No ripple in passband and have ripple in stopband. Group delay is nonlinear function of the frequency. Roll-off is high.

Elliptic Filters: Have ripple both in passband and stopband. Group delay is a nonlinear function of the frequency. Roll-off is the highest.

Bessel Filters: No ripple in passband and stopband. Group delay is constant. Roll-off is the lowest.

Fig. 4.30 The squared magnitude response of a typical Bessel filter

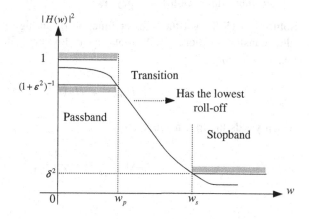

4.3.4 Analog Frequency Transformations

Once you have analogue low-pass prototype filter with cut-off frequency $w_c = 1$ rad/s, you can design other filters via frequency transformation. The possible frequency transformations are summarized as follows:

Lowpass to lowpass $s \leftarrow \dfrac{s}{w_c}$ where w_c is the desired cut$-$off frequency.

Lowpass to highpass $s \leftarrow \dfrac{w_c}{s}$ where w_c is the desired cut$-$off frequency.

Lowpass to bandpass $s \leftarrow \dfrac{s^2 + w_{c_l} w_{c_u}}{s(w_{c_u} - w_{c_l})}.$

Lowpass to bandpass $s \leftarrow \dfrac{s^2 + w_{c_l} w_{c_u}}{s(w_{c_u} - w_{c_l})}.$

Lowpass to bandpass $s \leftarrow \dfrac{s(w_{c_u} - w_{c_l})}{s^2 + w_{c_l} w_{c_u}}.$

w_{c_l} is the lower cut-off frequence.

w_{c_u} is the upper cut-off frequency.

Example 4.15 The transfer function of a low-pass analog filter with cut-off frequency $w_c = 1$ rad/s is given as

$$H_{lp}(s) = \frac{1}{(s+1)(s^2+s+1)}.$$

Using the above transfer function, find the transfer function of an high-pass analog filter with cut-off frequency $w_c = 1$ rad/s.

Solution 4.15 To get the transfer function of an high-pass filter from a low-pass filter transfer function, simply replace s in low-pass filter transfer function by $\frac{w_c}{s}$, i.e., $s \leftarrow \frac{w_c}{s}$, that is

$$H_{hp}(s) = H_{lp}(r)\big|_{r=\frac{w_c}{s}}$$

which yields the transfer function

$$H_{hp}(s) = \frac{1}{\left(\frac{1}{s}+1\right)\left(\frac{1}{s^2}+\frac{1}{s}+1\right)}$$

whose simplified form can be calculated as

$$H_{hp}(s) = \frac{s^2}{s^2 + s + 1} \frac{s}{s + 1}.$$

As it is seen from the above equation, the transfer function of a high pass filter includes s^i like terms in the numerator.

4.4 Implementation of Analog Filters

4.4.1 Low Pass Filter Circuits

Remember that the transfer function of the low-pass Butterworth filter was in the form

$$H(s) = \frac{w_c^N}{\prod_{k=1}^{N}(s - p_k)}. \qquad (4.102)$$

Considering (4.102), we can calculate the transfer function of the Butterworth filter for $w_c = 1$ and $N = 3$ as

$$H(s) = \frac{1}{(s + 1)(s^2 + s + 1)}. \qquad (4.103)$$

As it is also seen in (4.103), we can say that the transfer function of a low-pass filter has a constant number in its numerator, and at the denominator, we can have two different types of polynomials which are

$$(s + a) \quad (s^2 + b_1 s + b_2).$$

If we know how to implement $(s + a)$ and $(s^2 + b_1 s + b_2)$, then we can implement the transfer function $H(s)$ using circuit elements.

How to implement $H(s) = a/(s + a)$:

The transfer function $H(s) = a/(s + a)$ can be implemented using the circuit in Fig. 4.31.

Fig. 4.31 Analog implementation of $H(s) = a/(s + a)$ by circuit elements

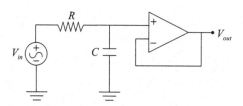

The transfer function of the circuit in Fig. 4.31 can be calculated as

$$H(s) = \frac{V_{out}(s)}{V_{in}(s)} \rightarrow H(s) = \frac{\frac{1}{RC}}{s + \frac{1}{RC}}.$$

How to implement $H(s) = b/(s+a)$:
The transfer function

$$H(s) = \frac{b}{s+a} \tag{4.104}$$

can be implemented using the circuit in Fig. 4.32.
The transfer function of the above circuit is

$$H(s) = \frac{V_{out}(s)}{V_{in}(s)} \rightarrow H(s) = \left(1 + \frac{R_3}{R_2}\right) \frac{\frac{1}{R_1 C}}{s + \frac{1}{R_1 C}}$$

How to implement $H(s) = a/s^2 + b_1 s + b_2$:
The transfer function

$$H(s) = \frac{a}{s^2 + b_1 s + b_2} \tag{4.105}$$

can be implemented using the circuit in Fig. 4.33.
The transfer function of the circuit in Fig. 4.33 can be calculated as

$$H(s) = \frac{V_{out}(s)}{V_{in}(s)} \rightarrow H(s) = \frac{\frac{K}{\tau_1 \tau_2}}{s^2 + \left(\frac{1}{\tau_1} + \frac{1}{R_2 C_1} + \frac{1-K}{\tau_2}\right)s + \frac{1}{\tau_1 \tau_2}}$$

where $K = 1 + R_B/R_A$, $\tau_1 = R_1 C_1$, $\tau_2 = R_2 C_2$. If common values are selected for the resistors R_1, R_2 and capacitors C_1, C_2, transfer function expression reduces to

Fig. 4.32 Analog implementation of $H(s) = b/(s+a)$ by circuit elements

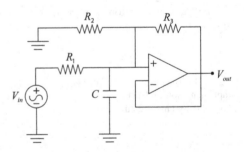

Fig. 4.33 Analog implementation of $H(s) = a/s^2 + b_1 s + b_2$ by circuit elements

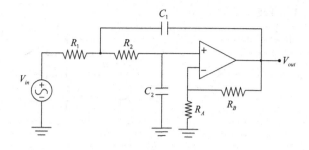

Fig. 4.34 Alternative analog implementation of (4.105)

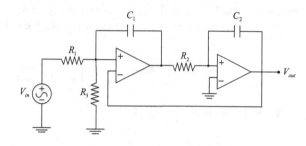

$$H(s) = K \frac{\frac{1}{\tau^2}}{s^2 + \frac{3-K}{\tau} s + \frac{1}{\tau^2}}$$

where $\tau = RC$.

An alternative implementation of (4.105) can be achieved using the circuit in Fig. 4.34.

The transfer function of the circuit in Fig. 4.34 can be calculated as

$$H(s) = \frac{\frac{1}{\tau_1 \tau_2}}{s^2 + \frac{1}{\tau_2} s + \frac{1 + R_1/R_3}{\tau_1 \tau_2}} \tag{4.106}$$

where $\tau_1 = R_1 C_1$, $\tau_2 = R_2 C_2$. If R_1 and R_3 are chosen as $R_1 = R_3$, then we get

$$H(s) = \frac{\frac{1}{\tau_1 \tau_2}}{s^2 + \frac{1}{\tau_2} s + \frac{2}{\tau_1 \tau_2}}. \tag{4.107}$$

Example 4.16 The transfer function of second order low-pass Butterworth filter with cut-off frequency $w_c = 1000$ rad/s is given as

$$H(s) = \frac{10^6}{s^2 + 1414s + 2 \times 10^6}.$$

Implement the given filter transfer function using circuit elements.

Fig. 4.35 Second order
low-pass filter
implementation

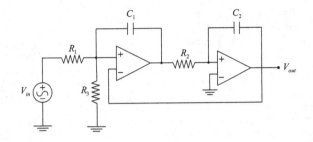

Solution 4.16 Let's use the circuit given in Fig. 4.35.

The transfer function of the circuit in Fig. 4.35 can be calculated as

$$H(s) = \frac{\frac{1}{\tau_1 \tau_2}}{s^2 + \frac{1}{\tau_2}s + \frac{1+R_1/R_3}{\tau_1 \tau_2}}. \tag{4.108}$$

When (4.38) is compared to

$$H(s) = \frac{10^6}{s^2 + 1414s + 2 \times 10^6}$$

we see that

$$\frac{1}{\tau_1 \tau_2} = 10^6 \qquad \frac{1}{\tau_2} = 1414 \qquad \frac{1+R_1/R_3}{\tau_1 \tau_2} = 2 \times 10^6. \tag{4.109}$$

In (4.109) let's first solve

$$\frac{1}{\tau_2} = 1414.$$

Since $\tau_2 = R_2 C_2$, if C_2 is chosen as 0.47 μF, then

$$R_2 = \frac{1}{1414 \times 0.47 \times 10^{-6}} \rightarrow R_2 = 1504 \ \Omega.$$

Next solving

$$\frac{1}{\tau_1 \tau_2} = 2 \times 10^6 \qquad \frac{1}{\tau_2} = 1414$$

for τ_1, we get $\tau_1 = 1414/2 \times 10^6$ and if C_1 is chosen as 0.47 µF, then

$$R_1 = 2 \times \frac{1414}{0.47} \rightarrow R_1 = 6017 \ \Omega.$$

Finally solving the equation

$$\frac{1 + R_1/R_3}{\tau_1 \tau_2} = 2 \times 10^6$$

for

$$\frac{1}{\tau_1 \tau_2} = 10^6$$

and

$$R_1 = 6017 \ \Omega$$

we find R_3 as

$$R_3 = R_1 = 6017 \ \Omega.$$

With the found values, our second order Butterworth low-pass filter circuit with cut-off frequency $w_c = 1000$ rad/s becomes as in Fig. 4.36.

The circuit in Fig. 4.36 includes some resistor values which may not be commercially available. In this case, we should use a resistor value closest to the calculated value in the Figure. This may slightly affect the accuracy of the filter. We can use the standard resistor and capacitor values shown in Tables 4.1 and 4.2. And to get the resistor value 6017 Ω in our example, we can use 6.2 KΩ or 5.6 KΩ and 430 Ω in series.

Fig. 4.36 Butterworth low-pass filter circuit with cut-off frequency $w_c = 1000$ rad/s

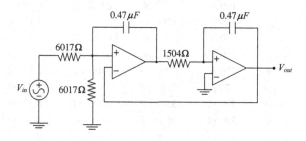

Table 4.1 Common resistor values for electronic circuits

Standard resistor values (±5%)							
1.0	10	100	1.0 K	10 K	100 K	1.0 M	10 M
1.1	11	110	1.1 K	11 K	110 K	1.1 M	11 M
1.2	12	120	1.2 K	12 K	120 K	1.2 M	12 M
1.3	13	130	1.3 K	13 K	130 K	1.3 M	13 M
1.5	15	150	1.5 K	15 K	150 K	1.5 M	15 M
1.6	16	160	1.6 K	16 K	160 K	1.6 M	16 M
1.8	18	180	1.8 K	18 K	180 K	1.8 M	18 M
2.0	20	200	2.0 K	20 K	200 K	2.0 M	20 M
2.2	22	220	2.2 K	22 K	220 K	2.2 M	22 M
2.4	24	240	2.4 K	24 K	240 K	2.4 M	
2.7	27	270	2.7 K	27 K	270 K	2.7 M	
3.0	30	300	3.0 K	30 K	300 K	3.0 M	
3.3	33	330	3.3 K	33 K	330 K	3.3 M	
3.6	36	360	3.6 K	36 K	360 K	3.6 M	
3.9	39	390	3.9 K	39 K	390 K	3.9 M	
4.3	43	430	4.3 K	43 K	430 K	4.3 M	
4.7	47	470	4.7 K	47 K	470 K	4.7 M	
5.1	51	510	5.1 K	51 K	510 K	5.1 M	
5.6	56	560	5.6 K	56 K	560 K	5.6 M	
6.2	62	620	6.2 K	62 K	620 K	6.2 M	
6.8	68	680	6.8 K	68 K	680 K	6.8 M	
7.5	75	750	7.5 K	75 K	750 K	7.5 M	
8.2	82	820	8.2 K	82 K	820 K	8.2 M	
9.1	91	910	9.1 K	91 K	910 K	9.1 M	

Table 4.2 Common capacitor values for electronic circuits

Standard capacitor values (±10%)						
10 pF	100 pF	1000 pF	0.010 mF	0.10 mF	1.0 mF	10 mF
12 pF	120 pF	1200 pF	0.012 mF	0.12 mF	1.2 mF	
15 pF	150 pF	1500 pF	0.015 mF	0.15 mF	1.5 mF	
18 pF	180 pF	1800 pF	0.018 mF	0.18 mF	1.8 mF	
22 pF	220 pF	2200 pF	0.022 mF	0.22 mF	2.2 mF	22 mF
27 pF	270 pF	2700 pF	0.027 mF	0.27 mF	2.7 mF	
33 pF	330 pF	3300 pF	0.033 mF	0.33 mF	3.3 mF	33 mF
39 pF	390 pF	3900 pF	0.039 mF	0.39 mF	3.9 mF	
47 pF	470 pF	4700 pF	0.047 mF	0.47 mF	4.7 mF	47 μF
56 pF	560 pF	5600 pF	0.056 mF	0.56 mF	5.6 mF	
68 pF	680 pF	6800 pF	0.068 mF	0.68 mF	6.8 mF	
82 pF	820 pF	8200 pF	0.082 mF	0.82 mF	8.2 mF	

4.4.2   Analog High-Pass Filter Circuit Design

Let's consider the transfer function of a high pass Butterworth filter given as

$$H_{hp}(s) = \frac{s^2}{s^2 + s + 1} \frac{s}{s + 1}. \tag{4.110}$$

Inspecting (4.110), we can conclude that the transfer function of a high pass filter contains two different terms

$$\frac{Ks^2}{s^2 + b_1 s + b_0}, \quad \frac{as}{s + b}. \tag{4.111}$$

Then if we know how to implement the terms in (4.111) by circuit elements, then we can construct a circuit for any high pass filter.

The high pass filter circuit can be obtained from a low pass filter circuit by replacing the resistors of the low pass filter by capacitors and replacing the capacitors of the low pass filter by resistors.

How to implement $H(s) = as/(s + b)$:

We can use the circuit in Fig. 4.37 to implement the transfer function

$$H(s) = \frac{as}{s + b}.$$

The transfer function of the circuit in Fig. 4.37 can be calculated in 's' domain. The transfer function of the circuit in Fig. 4.37 can be calculated as

$$H(s) = K \frac{s}{s + \frac{1}{R_1 C_1}} \quad \text{where} \quad K = 1 + \frac{R_2}{R_3}$$

If the resistors R_2 and R_3 are not used in Fig. 4.37, then the transfer function reduces to

$$H(s) = \frac{s}{s + \frac{1}{R_1 C_1}}. \tag{4.112}$$

Fig. 4.37 Analog implementation of $H(s) = as/(s + b)$ by circuit elements

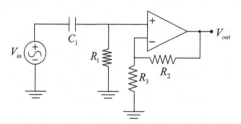

Fig. 4.38 Analog implementation of $H(s) = Ks^2/s^2 + b_1s + b_0$ by circuit elements

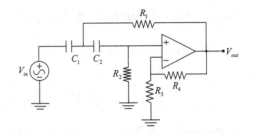

How to implement $H(s) = Ks^2/s^2 + b_1s + b_0$:

We can use the circuit in Fig. 4.38 to implement the transfer function

$$H(s) = \frac{Ks^2}{s^2 + b_1s + b_0}. \tag{4.113}$$

The circuit in Fig. 4.38 is called Sallen-Key topology whose transfer function can be calculated as

$$H(s) = \frac{Ks^2}{s^2 + \left(\frac{1}{\tau_2} + \frac{1}{R_2C_1} + \frac{1-K}{\tau_1}\right)s + \frac{1}{\tau_1\tau_2}} \tag{4.114}$$

where $\tau_1 = R_1C_1, \tau_2 = R_2C_2, K = 1 + R_4/R_3$.

If $R_1 = R_2$ and $C_1 = C_2$, then (4.114) reduces to

$$H(s) = \frac{Ks^2}{s^2 + \frac{3-K}{RC}s + \frac{1}{R^2C^2}}. \tag{4.115}$$

Example 4.17 Implement the high pass filter transfer function

$$H(s) = \frac{2.6s^2}{s^2 + 5.31s + 176.83}. \tag{4.116}$$

Solution 4.17 If we compare the given transfer function in (4.116) to

$$H(s) = \frac{Ks^2}{s^2 + \frac{3-K}{RC}s + \frac{1}{R^2C^2}}$$

we see that

$$\frac{2.6s^2}{s^2 + 5.31s + 176.83} = \frac{Ks^2}{s^2 + \frac{3-K}{RC}s + \frac{1}{R^2C^2}}$$

Fig. 4.39 High pass filter
circuit for Example 4.17

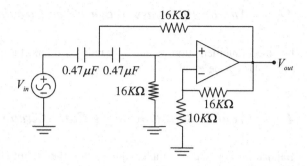

where we have

$$\frac{1}{R^2 C^2} = 176.83$$

And if we choose $C = 0.47$ μF, then R is found as

$$\frac{1}{R^2 (0.47 \times 10^{-6})^2} = 176.83 \rightarrow R^2 = \frac{10^{12}}{0.47^2 \times 176.83} \rightarrow R = 16000\ \Omega.$$

Also we have $K = 2.6$ since, $K = 1 + R_4/R_3$, we get

$$2.6 = 1 + \frac{R_4}{R_3} \rightarrow \frac{R_4}{R_3} = 1.6$$

Since $\frac{R_4}{R_3} = 1.6$, we can choose $R_4 = 16$ KΩ, $R_3 = 10$ KΩ. Then our high pass
filter circuit becomes as in Fig. 4.39.

Example 4.18 Implement high pass filter transfer function

$$H(s) = \frac{2.6 \times 0.5 s^2}{s^2 + 5.31 s + 176.83} \tag{4.117}$$

using circuit elements.

Solution 4.18 In (4.117); we have 0.5 factor in the numerator, for this reason we
add a voltage divider circuit to the end of the circuit which is shown in shadow in
Fig. 4.40.

4.4.3 Analog Bandpass Active Filter Circuits

For the implementation of analog bandpass filters, the prototype circuit shown in Fig. 4.41 can be employed.

4.4.4 Analog Bandstop Active Filter Circuits

Bandstop filters can be implemented using the circuit shown in Fig. 4.42.

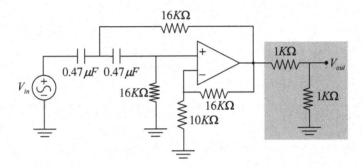

Fig. 4.40 High pass filter circuit with voltage divider

Fig. 4.41 Bandpass filter circuit

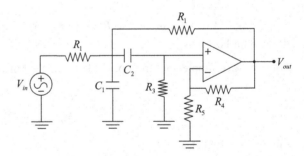

Fig. 4.42 Bandstop filter circuit

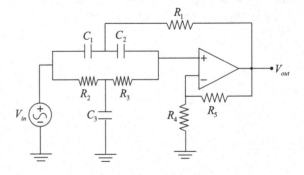

4.5 Infinite Impulse Response (IIR) Digital Filter Design (Low Pass)

Two methods are followed for the design of infinite impulse response digital filters, i.e., IIR filters. These methods are:

(1) Design an analog filter and convert it to a digital filter via sampling operation, i.e., digitize the designed analog filter to get the digital filter.
(2) Design the IIR digital filter directly.

We will use the first approach in this book. The steps for the design of IIR filters using analog prototypes are outlined in the Table 4.3.

Example 4.19 The magnitude response of a digital filter is depicted in the Fig. 4.43.

(a) By mapping the digital filter specifications to a continuous time, determine the continuous time filter specifications.
(b) Determine the squared magnitude response of the continuous time filter.

Solution 4.19 We will use the bilinear transformation method to find the digital filter specifications. In bilinear transformation, the relationship between analog and digital frequencies is

$$w_a = \frac{2}{T_s} \tan\left(\frac{w_d}{2}\right)$$

which can also be written as

$$w_d = 2\tan^{-1}\left(w_a \frac{T_s}{2}\right).$$

Table 4.3 Steps for the design of an IIR digital filter	IIR digital filter design using analog prototypes
	(1) Determine the digital filter specifications, such as w_p, w_s, R_p, R_s
	(2) Map digital filter frequency specifications to continuous time filter frequency specifications using a transformation method, for instance "bilinear transformation"
	(3) Design the continuous time filter according to continuous time specifications
	(4) Transform continuous time filter to digital filter using a transformation method, for instance "bilinear transformation"
	(5) Implement your digital filter by either designing a hardware using digital gates, or writing a software for digital devices which can be microprocessors, digital signal processing chips, or field programmable gate arrays (FPGA)

Fig. 4.43 The magnitude response of a digital filter

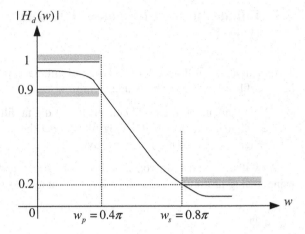

Since digital filter specifications are given, we should use

$$w_a = \frac{2}{T_s} \tan\left(\frac{w_d}{2}\right)$$

to find the analog filter specifications. Let $T_s = \frac{1}{2000}$ s, then the analog pass and stop frequencies are calculated as

$$w_{ap} = 4000 \tan\left(\frac{0.4\pi}{2}\right) \rightarrow w_{ap} = 2906.2 \text{ rad/s} \rightarrow w_{ap} = 925.54\pi$$

$$w_{as} = 4000 \tan\left(\frac{0.8\pi}{2}\right) \rightarrow w_{as} = 12{,}311 \text{ rad/s} \rightarrow w_{as} = 3918.7\pi.$$

Then the analog filter magnitude response can be drawn as in Fig. 4.44.

Fig. 4.44 Analog filter magnitude response

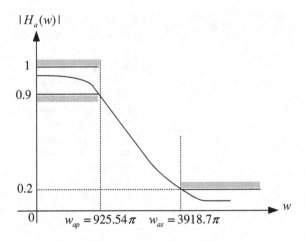

Using Fig. 4.44 the squared magnitude response of the analog filter can be found as in Fig. 4.45.

Example 4.20 The magnitude response of a lowpass digital filter is depicted in Fig. 4.46. State the digital filter specifications via mathematical expressions.

Solution 4.20 Since Fourier transform of the digital signals is periodic with period 2π, we can express the filter specifications for the interval $-\pi \leq w < \pi$. In addition, we know that aliasing in Fourier transform of a digital signal does not occur if magnitude response has nonzero values only for the interval $-\pi \leq w < \pi$.

For this reason, for the digital filters, we will only consider the frequency interval $-\pi \leq w < \pi$. In addition, the frequency interval $0 \leq |w| < \pi/2$ is accepted as the low frequency region and the frequency range $\pi/2 \leq |w| < \pi$ is accepted as the high frequency interval.

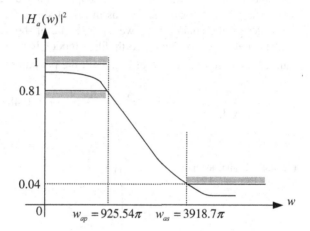

Fig. 4.45 Squared magnitude response of the analog filter

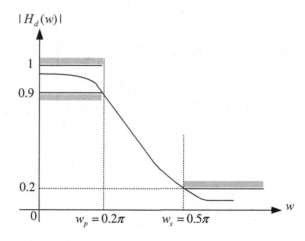

Fig. 4.46 The magnitude response of a lowpass digital filter

Then considering Fig. 4.46, the filter response can be expressed as

$$0.9 \leq |H_d(w)| \leq 1 \quad 0 \leq |w| \leq 0.2\pi,$$
$$|H_d(w)| \leq 0.2 \quad 0.5\pi \leq |w| \leq \pi.$$

Example 4.21 Design the digital filter with the following specifications

$$0.9 \leq |H_d(w)| \leq 1 \quad 0 \leq |w| \leq 0.4\pi,$$
$$|H_d(w)| \leq 0.2 \quad 0.8\pi \leq |w| \leq \pi.$$

Solution 4.21 Using the given specifications we can draw the magnitude response of the digital filter as in Fig. 4.47.

For the design of our digital filter, we first convert digital filter specifications to analog filter specification using the bilinear transformation method. Since this example is a continuation of Example 4.19, we can use the converted parameters of Example 4.19. Using the results of Example 4.19, we can analog draw the analog filter squared magnitude response as in Fig. 4.48.

To design the analog filter, we can use one of the available analog prototypes models. Let's choose Butterworth filter model for our design. From the given squared magnitude response in Fig. 4.48, the parameters ϵ^2, ϵ and δ^2 are found as

$$\frac{1}{\sqrt{1+\epsilon^2}} = 0.81 \rightarrow \epsilon^2 = 0.2346 \rightarrow \epsilon = 0.4843 \quad \delta^2 = 0.04.$$

Fig. 4.47 Digital lowpass filter for Example 4.21

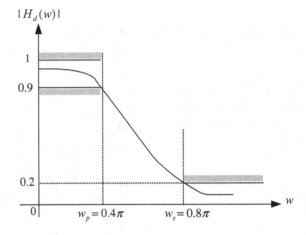

Fig. 4.48 Squared magnitude response of the analog filter obtained from digital filter specifications after bilinear transformation operation

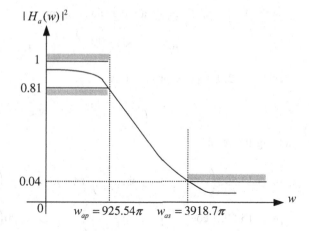

The parameters $1/d$ and $1/k$ are calculated as follows

$$\frac{1}{d} = \sqrt{\frac{\delta^{-2} - 1}{\epsilon^2}} \rightarrow \frac{1}{d} = \sqrt{\frac{25 - 1}{0.2346}} \rightarrow \frac{1}{d} = 10.1144,$$

$$\frac{1}{k} = \frac{w_s}{w_p} \rightarrow \frac{1}{k} = 4.2340.$$

And the filter order is calculated as

$$N \geq \frac{\log\left(\frac{1}{d}\right)}{\log\left(\frac{1}{k}\right)} \rightarrow N \geq \frac{\log(10.1144)}{\log(4.234)} \rightarrow N \geq 1.6 \rightarrow N = 2.$$

The cutoff frequency is calculated via

$$w_p \epsilon^{-\frac{1}{N}} \leq w_c \leq w_s \left(\delta^{-2} - 1\right)^{-\frac{1}{2N}} \rightarrow 925.54\pi \times (0.4843)^{-\frac{1}{2}} \leq w_c \leq 3918.7\pi \times 24^{\frac{1}{4}}$$

leading to

$$1308\pi \leq w_c \leq 8619\pi. \tag{4.118}$$

And considering (4.118), we can choose w_c as

$$w_c = \frac{1308\pi + 8619\pi}{2} \rightarrow w_c = 4963\pi \rightarrow w_c = 15{,}592.$$

In the last step, the poles are calculated using

$$p_k = w_c e^{j\frac{\pi}{2}\left(1 + \left(\frac{2k-1}{N}\right)\right)}, \quad k = 1, \ldots, N.$$

For $N = 2$, the poles are found as

$$p_1 = w_c e^{j\frac{3\pi}{4}} \quad p_2 = w_c e^{j\frac{5\pi}{4}}$$

yielding the results

$$
\begin{aligned}
p_1 &= 15{,}592 \left(\cos\left(\frac{3\pi}{4}\right) + j\sin\left(\frac{3\pi}{4}\right) \right) \rightarrow p_1 = 7796\left(-\sqrt{2} + j\sqrt{2}\right), \\
p_2 &= 15{,}592 \left(\cos\left(\frac{5\pi}{4}\right) + j\sin\left(\frac{5\pi}{4}\right) \right) \rightarrow p_2 = 7796\left(-\sqrt{2} - j\sqrt{2}\right).
\end{aligned}
\tag{4.119}
$$

The transfer function is found using

$$H_a(s) = \frac{w_c^N}{(s - p_1)(s - p_2) \cdots (s - p_N)}. \tag{4.120}$$

Substituting the calculated poles in (4.119) into (4.120) for $N = 2$, we get

$$H_a(s) = \frac{15{,}592^2}{\left(s + 7796\sqrt{2} - j7796\sqrt{2}\right)\left(s + 7796\sqrt{2} + j7796\sqrt{2}\right)}$$

which is simplified as

$$H_a(s) = \frac{15{,}592^2}{\left(s + 7796\sqrt{2}\right)^2 + \left(7796\sqrt{2}\right)^2}$$

leading to the result

$$H_a(s) = \frac{243{,}110{,}464}{s^2 + 22{,}050s + 2.43 \times 10^8}.$$

We are done with the analog filter design. Since our aim was to design the digital filter, we should digitize our analog filter to find the digital filter. For this purpose, we will use bilinear transformation method. The conversion procedure is outlined as:

$$H_d(z) = H_a(s)\big|_{s = \frac{2}{T_s}\frac{1 - z^{-1}}{1 + z^{-1}}} \tag{4.121}$$

Using $T_s = \frac{1}{2000}$ s in (4.121), we get

$$H_d(z) = \frac{243{,}110{,}464}{\left(4000\frac{1-z^{-1}}{1+z^{-1}}\right)^2 + 22{,}050\left(4000\frac{1-z^{-1}}{1+z^{-1}}\right) + 2.43 \times 10^8}. \tag{4.122}$$

When (4.122) is simplified, we obtain

$$H_d(z) = \frac{243{,}110{,}464 \times (1 + 2z^{-1} + z^{-2})}{10^7 \times (34.72 + 12.02z^{-1} + 1.6z^{-2})}$$

which can be rearranged as

$$H_d(z) = \frac{24.3 + 48.6z^{-1} + 24.3z^{-2}}{34.72 + 12.02z^{-1} + 1.6z^{-2}}.$$

To implement the digital filter with the above transfer function, we need to express the filter output-input relation in time domain. This is possible using

$$H_d(z) = \frac{Y(z)}{X(z)} \rightarrow \frac{Y(z)}{X(z)} = \frac{24.3 + 48.6z^{-1} + 24.3z^{-2}}{34.72 + 12.02z^{-1} + 1.6z^{-2}}$$

from which we get

$$34.72y[n] + 12.02y[n-1] + 1.6y[n-2] = 24.3x[n] + 48.6x[n-1] + 24.3x[n-2]$$

which leads to the expression

$$y[n] = -0.34y[n-1] - 0.05y[n-2] + 0.7x[n] + 1.4x[n-1] + 0.7x[n-2] \tag{4.123}$$

where $x[n]$ is the input of the digital filter and $y[n]$ is the filtered signal.

And the Eq. (4.123) can be implemented using a computer program, or the filter can be implemented in other digital hardware such as microprocessors, DSP chips, FPGAs, via hardware programming languages such as assembly, VHDL, etc., or an application specific digital hardware consisting of gates and other digital devices can be specifically produced for this filter.

4.5.1   Generalized Linear Phase Systems

A LTI system is said to be a generalized linear phase system if its transfer function is of the form

$$H(w) = A_r(w)e^{-j(\beta + \alpha w)} \tag{4.124}$$

where $A_r(w)$ is a real function of w. Considering (4.124), the group delay is calculated as

$$\tau_g(w) = -\frac{d\theta(w)}{dw} \rightarrow \tau_g(w) = \alpha. \tag{4.125}$$

A causal LTI system is a linear phase system if its $L+1$ point impulse response $h[n]$ satisfies

$$h[n] = \pm h[L-n] \quad 0 \le n \le L \tag{4.126}$$

where L can be an odd or even integer. And for such systems, the Fourier transform of $h[n]$ happens to be in the form

$$H(w) = A_r(w)e^{-\frac{jwL}{2}}. \tag{4.127}$$

4.6 Finite Impulse Response (FIR) Digital Filter Design

In many practical applications, FIR filters are preferred over their IIR counterparts. The main advantages of FIR filter over IIR filter can be summarized as follows:

(1) Most IIR filters have nonlinear phase characteristics, which creates problem for practical applications.
(2) FIR filters having linear phase responses and they can be easily designed.
(3) FIR filters can be implemented efficiently with affordable computational overhead.
(4) Stable FIR filters can be designed in an easy manner.
(5) In the literature, there exist excellent FIR filter design techniques.

The main disadvantage of the FIR filters over IIR filters is that for the applications requiring narrow band transitions, i.e. steep roll-off, more arithmetic operations are required which means that more digital hardware components such as adders, multiplexers, multipliers, etc., are required.

Designing FIR filter is nothing but determining the impulse response of an LTI system. The impulse response of the LTI system under concern includes a finite number of samples. If $h[n]$ denotes the impulse response of a FIR filter, then the output of the filter is written as:

$$y[n] = \sum_{k=-L}^{M} h[k]x[n-k]$$

where usually $L = M$ is assumed. If $h[n] = 0$ for $n < 0$, then the filter is said to be a causal filter. Otherwise, we have an anti-causal filter. Causal filters are practically realizable; on the other hand, anti-causal filters cannot be implemented. For this reason, anti-causal FIR filters should be transferred to causal FIR filters to enable their use in practical systems.

4.6.1   FIR Filter Design Techniques

There are basically three methods used for the design of FIR filters. These methods are

(a) FIR filter design by windowing.
(b) FIR filter design by frequency sampling.
(c) Equiripple FIR filter design.

Now let's see the first method.

4.6.1.1   FIR Filter Design by Windowing

Design of FIR Filter in Time Domain:

The frequency response of an ideal low pass digital filter is shown in the Fig. 4.49 where only one period of the frequency response around origin is depicted.

And we know that $H_{id}(w)$ satisfies $H_{id}(w) = H_{id}(w + m2\pi)$. The time domain expression for the low pass digital filter can be calculated as

$$h_{id}[n] = \frac{1}{2\pi} \int_{-w_c}^{w_c} \underbrace{H_{i_{lp}}(w)}_{=1} \times e^{jwn} dw$$

$$= \frac{1}{\pi n} \sin(w_c n) \quad n = 0, \pm 1, \pm 2, \ldots$$

where w_c is called cut-off frequency. It is clear that $h_{id}[n]$ includes an infinite number of samples. And the convolutional operation cannot be realized using an

Fig. 4.49 The frequency response of an ideal low pass digital filter

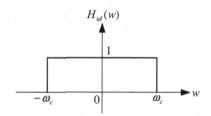

infinite number of samples. To alleviate this obstacle, we truncate the ideal filter and obtain the FIR filter as

$$h[n] = \begin{cases} h_{id}[n] & if \ |n| \leq L \\ 0 & otherwise \end{cases}$$

which can also be written as

$$h[n] = h_{id}[n] \times w[n]$$

where $w[n]$ is the rectangular window defined as

$$w[n] = \begin{cases} 1 & if \ |n| \leq L \\ 0 & otherwise. \end{cases}$$

This type of design approach is straightforward. However, such a designed filter suffers from Gibbs phenomenon. In addition, since the used window is anti-causal so is the FIR filter. However, we can obtain a causal window via truncation as follows

$$w[n] = \begin{cases} 1 & if \ 0 \leq n \leq L \\ 0 & otherwise. \end{cases} \tag{4.128}$$

To alleviate the effects of Gibbs phenomenon, windows other than the rectangular one such as, Hamming, Hanning, Bartlett, Triangular, and Blackman are used.

Design of FIR Filter in Frequency Domain:

Assume that $H(w)$ is the frequency response of a FIR filter in a way that it minimizes the error

$$\epsilon = \frac{1}{2\pi} \int\limits_{-\pi}^{\pi} |H(w) - H_{id}(w)|^2 dw$$

where applying the Parseval's identity, we get

$$\epsilon = \sum_{n=-\infty}^{\infty} |h[n] - h_{id}[n]|^2 \rightarrow$$

$$\epsilon = \sum_{n=0}^{L} |h[n] - h_{id}[n]|^2 + \sum_{n=Z-[0\ L]} |h[n] - h_{id}[n]|^2. \tag{4.129}$$

When (4.129) is equated to zero, we obtain

$$h[n] = \begin{cases} h_{id}[n] & \text{if } 0 \leq n \leq L \\ 0 & \text{otherwise.} \end{cases}$$

Properties of Windows:

Let $W_n(w)$ be the frequency response of the window. The main-lobe of the window is defined as the region between the first zero crossings on the left and right sides of the origin.

The width of the main-lobe of the causal rectangular window is approximated as

$$\Delta w = \frac{4\pi}{L+1}. \tag{4.130}$$

It is desirable to have a main lobe as narrow as possible. The width of the main-lobe controls the amount of attenuation on passband region. Side-lobes are the regions extending from first zero crossings points on either side of the origin.

Side-lobes are responsible for the ripples occurring in passband and stopband. For a wide range of frequencies, pass and stop band ripples are equal to each other.

For the causal rectangular window increasing the window length L, decreases the width of the main-lobe, however the areas under side-lobes stays the same which means that ripples occurs with the same amplitude but more frequently. To reduce the amount of area under ripples or to reduce the height of the ripples; we need to rub the ends of the rectangular window for a smoother transition to zero.

For this purpose, we employ some commonly used windows as outlined below:

Hanning Window:

$$w[n] = \begin{cases} 0.5 - 0.5 \cos\left(\frac{2\pi n}{L}\right) & \text{if } 0 \leq n \leq L \\ 0 & \text{otherwise} \end{cases} \tag{4.131}$$

Hamming Window:

$$w[n] = \begin{cases} 0.54 - 0.46 \cos\left(\frac{2\pi n}{L}\right) & \text{if } 0 \leq n \leq L \\ 0 & \text{otherwise} \end{cases} \tag{4.132}$$

Blackman Window:

$$w[n] = \begin{cases} 0.42 - 0.5 \cos\left(\frac{2\pi n}{L}\right) + 0.08 \cos\left(\frac{4\pi n}{L}\right) & \text{if } 0 \leq n \leq L \\ 0 & \text{otherwise} \end{cases} \tag{4.133}$$

For the Hanning, Hamming, and Blackman windows the general form can be written as

$$w[n] = \begin{cases} a + b\cos\left(\frac{2\pi n}{L}\right) + c\cos\left(\frac{4\pi n}{L}\right) & \text{if } 0 \le n \le L \\ 0 & \text{otherwise} \end{cases} \quad (4.134)$$

where for Hanning window $a = 0.5, b = -0.46, c = 0$, and for Blackman window $a = 0.42, b = -0.5, c = 0.08$.

Bartlett (Triangular) Window:

$$w[n] = \begin{cases} \frac{2n}{L} & \text{if } 0 \le n \le \frac{L}{2} \\ 2 - \frac{2n}{L} & \text{if } \frac{L}{2} < n \le L \\ 0 & \text{otherwise} \end{cases} \quad (4.135)$$

In Table 4.4 five different windows are compared to each other considering mainlobe width and peak sidelobe amplitude.

All the windows given up to now can be approximated by the Kaiser window. Now let's give some information about Kaiser window.

Kaiser Window:

The Kaiser window is defined as

$$w[n] = \begin{cases} \dfrac{I_0\left[\beta\left(1 - \left[\frac{n-\alpha}{\alpha}\right]^2\right)^{\frac{1}{2}}\right]}{I_0(\beta)} & \text{if } 0 \le n \le L \\ 0 & \text{otherwise} \end{cases} \quad (4.136)$$

where $I_0(\cdot)$ is the modified Bessel function of the first kind which is equal to

$$I_0(x) = \frac{1}{2\pi} \int_0^{2\pi} e^{x\cos\theta} d\theta \quad (4.137)$$

and $\alpha = M/2$, β is the design parameter given by

$$\beta = \begin{cases} 0.1102(C - 8.7) & C > 50 \\ 0.5842(C - 21)^{0.4} + 0.07886(C - 21) & 21 \le C \le 50 \\ 0.0 & C < 21 \end{cases} \quad (4.138)$$

Table 4.4 Windows and their properties

Window type	Mainlobe width	Peak sidelobe amplitude (dB)
Rectangular	4 π/(2L + 1)	−13
Bartlett	8 π/L	−27
Hanning	8 π/L	−32
Hamming	8 π/L	−43
Blackman	12 π/L	−58

where the parameter C is defined as

$$C = -20\log_{10}\rho. \tag{4.139}$$

2ρ is the maximum ripple available in the passband. Let the transition region width be defined as $\Delta w = w_s - w_p$. With the given filter specifications, the order of the Kaiser window is found as

$$L = \frac{C-8}{2.285\Delta w} \tag{4.140}$$

which is also the length of the FIR filter satisfying the given specifications.

Example 4.22 Find the impulse response of a FIR filter whose specifications are given as

$$w_p = 0.4\pi \quad w_s = 0.8\pi \quad \rho = 0.01.$$

Solution 4.22 First we need to calculate the order of the Kaiser window given as

$$L = \frac{C-8}{2.285\Delta w}$$

where the parameters are calculated as

$$\Delta w = w_s - w_p \rightarrow \Delta w = 0.8\pi - 0.4\pi \rightarrow \Delta w = 0.4\pi$$
$$C = -20\log_{10}\rho \rightarrow C = -20\log_{10}0.01 \rightarrow C = 40$$

And the length of the window is found as

$$L = \frac{C-8}{2.285\Delta w} \rightarrow L = \frac{40-8}{2.285 \times 0.4\pi} \rightarrow L = 12$$

Next, we calculate the design parameter β as follows

$$\beta = 0.5842(C-21)^{0.4} + 0.07886(C-21) \rightarrow$$
$$\beta = 0.5842(40-21)^{0.4} + 0.07886(40-21) \rightarrow \beta = 3.3953$$

The function $I_0(\beta)$ can be approximated as

$$I_0(\beta) \approx 1 + \frac{\beta^2}{2} + \frac{\beta^4}{64} + \frac{\beta^6}{2304} + \frac{\beta^8}{147,456}$$

or we need to write a computer program for the computation of the integral expression in (4.137). Using the definition of $w[n]$

$$w[n] = \begin{cases} \dfrac{I_0\left[\beta\left(1-\left[\frac{n-\alpha}{\alpha}\right]^2\right)^{\frac{1}{2}}\right]}{I_0(\beta)} & 0 \le n \le L \\ 0 & otherwise \end{cases}$$

the window elements for $L = 12, \beta = 3.3953, \alpha = L/2$ can be calculated as

$$w[n] = [\underset{n=0}{\underbrace{0.15}} \quad 0.31 \quad 0.5 \quad 0.69 \quad 0.85 \quad 0.96 \quad 1 \quad 0.96 \quad 0.85 \quad 0.69 \quad 0.5 \quad 0.31 \quad 0.15].$$

And the FIR filter coefficients are evaluated using

$$h[n] = h_{id}[n]w[n]$$

where ideal filter coefficients are

$$h_{id}[n] = \frac{1}{\pi n}\sin(w_c n)$$

for which w_c can be calculated as

$$w_c = \frac{w_p + w_s}{2} \rightarrow w_c = 0.6\pi.$$

Hence, ideal filter coefficients can be calculated as

$$h_{id} = [\underset{n=0}{\underbrace{0.6}} \quad 0.30 \quad -0.09 \quad -0.06 \quad 0.07 \quad 0 \quad -0.05 \quad 0.03 \quad 0.02 \quad -0.03$$
$$0 \quad 0.03 \quad -0.016].$$

Finally the FIR filter coefficients are found using

$$h[n] = h_{id}[n]w[n]$$

as

$$h[n] = [0.09 \quad 0.093 \quad -0.045 \quad -0.041 \quad 0.059 \quad 0 \quad -0.05 \quad 0.029 \quad 0.017$$
$$-0.02 \quad 0 \quad 0.009 \quad -0.0024]$$

4.6.1.2 FIR Filter Design by Frequency Sampling

Let $H(w)$ be the Fourier transform of the impulse response of the FIR filter to be designed. If we take L samples from $H(w)$ via sampling operation as in

$$H[k] = H(w)|_{w=\frac{k2\pi}{L}} \quad k = 0, 1, \ldots, L-1 \tag{4.141}$$

we obtain the DFT coefficients $H[k]$. Using (4.141) in inverse DFT formula

$$h[n] = \frac{1}{L} \sum_{k=0}^{L-1} H[k] e^{jk\frac{2\pi}{L}}, \quad n = 0, 1, \cdots, L-1 \tag{4.142}$$

we obtain the impulse response of digital FIR filter.

4.7 Problems

(1) Convert the differential equation

$$\frac{d^2 y(t)}{dt^2} + 4\frac{dy(t)}{dt} + 3y(t) = \frac{dx(t)}{dt} - x(t)$$

to a difference equation via sampling operation and find the transfer function of the difference equation.

(2) For a continuous time LTI system, the relation between system input and system output is given via the differential equation

$$\frac{d^2 y(t)}{dt^2} + 2\frac{dy(t)}{dt} - 3y(t) = \frac{d^2 x(t)}{dt^2} + 2x(t).$$

Considering this LTI system:

(a) Find the transfer function $H(s)$ of the LTI system. Decide on whether the system has the stability property or not.
(b) Convert the transfer function to its discrete equivalent, for this purpose take the sampling period as $T_s = 1$.

(3) The specifications of a low-pass analog filter are given as

$$w_p = 1000 \text{ rad/san} \quad w_s = 8000 \text{ rad/san} \quad R_p = 10 \text{ dB} \quad R_s = 40 \text{ dB}.$$

(a) Find the transfer function $H(s)$ of this filter. In other words, design your analog filter with the given specifications in the problem. For your design, use Butterworth, Chebyshev Type-I, and Chebyshev Type-II filter design methods separately.
(b) Implement your filters using circuit elements.

(4) The specifications of a low-pass IIR digital filter are given as

$$w_p = 0.1\pi \,\text{rad/s} \quad w_s = 0.7\pi \,\text{rad/s} \quad R_p = 10 \,\text{dB} \quad R_s = 40 \,\text{dB}.$$

  (a) Find the transfer function $H(z)$ of this filter. Use sampling period $T_s = 1$ in your design.
  (b) Using $H(z)$, write a difference equation between filter input and filter output.

(5) Design the FIR digital filter whose specifications are given as

$$w_p = 0.4\pi \quad w_s = 0.8\pi \quad \rho = 0.01.$$

In your design use the windowing approach, and use Kaiser window for your design.

Bibliography

1. Discrete-Time Signal Processing by A. V. Oppenheim and R. W. Schafer.
2. Digital Signal Processing: Principles, Algorithms, and Applications by J. G. Proakis and D. G. Manolakis.
3. Digital Signal Processing in Communication Systems by Marvin E. Frerking.
4. Multirate Digital Signal Processing by R. E. Crochiere and L. R. Rabiner.
5. Digital Signal Processing by William D. Stanley.

© Springer Nature Singapore Pte Ltd. 2018
O. Gazi, *Understanding Digital Signal Processing*, Springer Topics
in Signal Processing 13, DOI 10.1007/978-981-10-4962-0

Index

© Springer Nature Singapore Pte Ltd. 2018
O. Gazi, *Understanding Digital Signal Processing*, Springer Topics
in Signal Processing 13, DOI 10.1007/978-981-10-4962-0

Printed in the United States
By Bookmasters